I0503782

Moubray's Treatise on Domestic and Ornamental Poultry

A Practical Guide to the History, Breeding, Rearing, Fattening and General Management of Fowls and Pigeons

by John Lawrence & L.A. Meall

with an introduction by Jackson Chambers

This work contains material that was originally published in 1854.

This publication is within the Public Domain.

This edition is reprinted for educational purposes and in accordance with all applicable Federal Laws.

Introduction Copyright 2018 by Jackson Chambers

COVER CREDITS

Wrap-Around Cover
The Poultry Farm (1652) by Adriaen van Utrecht
[Public Domain]
Photographed by *Handsaw* from a private collection,
via Wikimedia Commons

Research / Resources
Wikimedia Commons
www.Commons.Wikimedia.org

Many thanks to all the incredible photographers, artists,
researchers, biographers, historians, and archivists who share
their great work via the Wikipedia family.

PLEASE NOTE :
As with all reprinted books of this age that are intended to perfectly reproduce the original edition,
considerable pains and effort had to be undertaken to correct fading and sometimes outright damage to
existing proofs of this title. At times, this task can be quite monumental, requiring an almost total
rebuilding of some pages from digital proofs of multiple copies. Despite this, imperfections still sometimes
exist in the final proof and may detract slightly from the visual appearance of the text.

DISCLAIMER :
Due to the age of this book, some methods or practices may have been deemed unsafe or
unacceptable in the interim years. In utilizing the information herein, you do so at your
own risk. We republish antiquarian books without judgment or revisionism, solely
for their historical and cultural importance, and for educational purposes.

Self Reliance Books

Get more historic titles on animal and stock breeding, gardening and old fashioned skills by visiting us at:

http://selfreliancebooks.blogspot.com/

PREFACE.

No apology is offered for adding another to the already numerous volumes issued of late in the department of poultry literature : the sale of *nine* previous editions of the original Treatise is considered sufficient to justify the Publishers in producing a new and enlarged work, adapted to the requirements of the present day.

The value and practical utility of the Treatise is established by the fact, that its pages have been liberally and unscrupulously drawn upon by subsequent writers, who have gladly, though unthankfully, availed themselves of the great store of original information it contained.

Of his own labours (which have not been light) the Editor would only remark, that he has made no attempt, by the aid of fine writing, to invest the subject with an undue interest, his sole object having been to follow in the Author's steps, and to produce as useful and practical a work as possible, from which the novice, and perhaps the experienced amateur, may derive information and assistance in the pursuit of poultry-keeping. Discarding all preconceived notions and prejudices, he has *not been content to take anything upon trust*, merely because long popularly accepted; and every fact and statement capable of being investigated, has been subjected, whenever practicable, to the test of every-day experience. Whether he has succeeded in his task, the reader will be the best judge. It is a matter of regret with him, that much interesting matter has been necessarily omitted, as the insertion would have involved the extension of the volume to unreasonable limits.

Of the pictorial department of the work, the Editor may speak with more freedom and confidence. The portraits (nearly all from life) were drawn by Mr. C. J. Winter, of Great Yarmouth, and as far as he can judge, the Editor believes that the life, character, and truthfulness imparted to them, render them inferior to none that have yet appeared.

L. A. M.

Gt. Yarmouth,
June, 1854.

CONTENTS.

ERRATA.

Page 220, line 11. *For* Plate V. *read* Plate VI.

Page 233, line 15. *For* Plate V. *read* Plate VI.

LIST OF ILLUSTRATIONS.

Drawn by C. J. Winter.
Engraved and printed in colours by J. M. Kronheim & Co.

LIST OF ILLUSTRATIONS.

DOMESTIC POULTRY.

CHAPTER I.

PRIMITIVE DOMESTICATION AND EARLY NOTICES OF FOWLS.

WHATEVER diversity of opinion may have existed among those who have written upon our DOMESTIC POULTRY, as to their origin, natural history, or habits, there is one point, at least, on which they all seem to have agreed, namely, that it is utterly impossible to trace back the history of these fowls, so as satisfactorily to fix the period of their first domestication among the nations of the earth. And although we do not imagine that our efforts will be attended with much better success than has been attained by those who have preceded us, it may not prove uninteresting, nor, indeed, altogether unprofitable, to pursue the inquiry so far as is consistent with (what we trust will be found) the practical design and usefulness of our work, and the limits within which it is prescribed.

We have no history from which we can gather

any account of the first domestication of our fowls ; and, as might be supposed, the earliest notices of them are to be found in the volume of Holy Writ. A writer in the *Gardener's Chronicle* has hazarded an opinion, that the event is coeval with the keeping of sheep by Abel, and the tilling of the ground by Cain—a conjecture unwarranted even by probability : for it is hardly possible for the nomadic and wandering tribes of the earth to have accomplished their domestication—the very term almost presupposing the possession of fixed habitations and domestic abodes by those who first subjugated the " fowls of the air " for their own uses. Whenever it may have taken place it is not likely to have been the work of a day, or by any means an easy task : nor is it probable that it was completed, before a long series of gradual attempts and many fruitless efforts had been made.

From the injunction given to Abraham nearly 4000 years ago, and imposed by the Levitical law upon the Israelites, to sacrifice pigeons or turtle doves, it would seem that these birds were as much, if not more, accessible to the people generally, than the animals of their herds or flocks ; and not improbably were kept for sacrificial purposes. But there does not seem any ground for believing that an attempt was made at that early period to domesticate them for the purposes of food.

Some writers have selected the Babylonish Captivity (B. C. 500) as the date at which the domestication of poultry was first practised: but we are inclined to believe that we shall be nearer the truth if we go further back into remote antiquity, fixing upon the establishment of the Jewish kingdom by David, some 3000 years ago, at a time when the arts of husbandry and social life are most likely to have had their rise, and to have spread among the people. Thus the Psalmist, speaking of the period to which the date we allude to may be assigned, mentions "*feathered fowl*," or, as the Hebrew original has it, "*fowls of the wing;*" indicating such an acquaintance with those which are comprised in the order to which our poultry belong, as enabled him to distinguish between them and the "fowls of the wing." Shortly after, too, we read of King Solomon consuming daily in his household "fatted fowl:" and whilst we are not disposed to go the length of those who have rendered the Hebrew words (*barburim abusim*) as meaning capons, (of the very existence of which there is no evidence,) we are at least justified in considering the passage to prove tolerably satisfactorily the approximate correctness of our conjecture. Five hundred years later we also find it recorded that the prophet Nehemiah had daily prepared for his household, "an ox, six choice sheep, and fowls." About this time also Bishop Pococke tells us, the Bey of Tunis daily partook, at his meals, of both "fowls and eggs."

The observance of the Levitical law must have caused a large consumption of PIGEONS and Doves by the Israelitish people, and we may hence infer that those birds were among the earliest domestic-ated. From a passage in Isaiah (ch. lx. ver. 8) it is evident that at a period certainly 600 years before the Christian era, they were very generally kept in cotes, or houses with " windows." Indeed the rearing and selling of pigeons must have become an important branch of traffic even at Jerusalem; for when our Saviour visited the temple, we read that he overthrew the seats of " *those that sold doves therein*." The domestication of pigeons among the ancient Pagan nations appears to have been scarcely less remote than among the Jews: of this abundant evidence is found in the works of Æsop, Anacreon, Virgil, and others. In the pages of those writers, frequent allusion is made to the bird as a " Bearer of Epistles," or Carrier: thus Ovid tells us, that Taurosthenes, by means of a pigeon stained with a purple dye, conveyed to his father at Ægina information of his victories in the Olympic games, on the same day on which they occurred. In a similar manner, also, did the besieged at Modena keep up a communication with their approaching allies when every other channel was closed; of this Pliny speaks when he asks,— " What availed Antony all his cares, when the messenger was cleaving the air?"

The flesh of the SWAN was, by the law of Moses,

forbidden to be eaten as unclean; and under this prohibition was most likely included that of all other web-footed fowls.

The PEACOCK is mentioned by Job, (B. C. 1520,) and in 1 Kings, ch. x. ver. 22: we are there told that these birds were brought to King Solomon by the navy of Tharshish, (B. C. 1000,) together with ivory and apes " from Ophir ; " which has been variously conjectured to be Africa, Tunis, or Arabia. Buffon erroneously (as we think) asserts that the peacock was introduced into Greece, and thence into Europe, by Alexander the Great (B. C. 330) : though this is certainly incorrect, as the peacock is mentioned by Eupolis, Aristophanes, Antiphon, and other classic writers who flourished a century previous to that period. By the last-named author we are told that in his day a pair of pea-fowls were valued at 1000 drachmæ, (a sum equal in value to £30 of our money,) and that they were publicly exhibited at Athens as a spectacle. That the pea-fowl was common, or at least generally known, in the time of Alexander, is clear from the way in which Aristotle compares vain and jealous persons to the peacock—evidently selecting a familiar illustration of the foibles he censured. We also gather from the writings of Cicero and others, that this bird formed an almost indispensable addition at a Roman banquet—the brains being especially esteemed by the epicures: and hence those fowls were regularly reared and fattened for the table.

Although we meet with no direct allusion to GEESE or DUCKS in Holy Writ, there is little doubt, as we have before observed, that they were included generally among the unclean fowls; as the frequently recurring figurations of those birds among the hieroglyphical works of the Egyptians preclude the idea that they were unknown to the ancients. So far, indeed, from that being the case, we are assured by Herodotus (B. C. 430) that long before his time quails and ducks formed the customary food of that people.

GEESE were probably domesticated much earlier than ducks. Homer (about eight centuries before Christ) notices them, and describes their being fed upon grain steeped in water. Æsop (B. C. 600) imagines a man bringing up a swan and a goose together, " the one for his ear, and the other for his belly." Among the Romans it was dedicated to Juno : and it is said that one of its kind, by its cackling, saved the imperial city (B. C. 388) from an enemy,—a story that every school-boy is acquainted with, and which has been transmitted to us by Livy, Lucretius, Virgil, and others, who, with more than ordinary care and ampleness of detail, inform us that it was a white or silver goose to whom the Romans were indebted for the signal service referred to. Gratitude for their usefulness on that occasion did not, however, according to Pliny, prevent the Roman gourmands from subjecting them to the cruel process of " cramming;" invented for the express purpose of furnishing the dainty dish of diseased or enlarged livers. The same author also records that when the Romans overran the Netherlands, (B. C. 57,) they occupied much of their time in catching geese (wild, of course, we infer) for their flesh and feathers,—the former, when roasted, being so much esteemed by the soldiers as food, and the latter being no less sought after in order to provide the means of resting softly at night.

From the directions given by Columella (A. D. 40) for rearing geese and ducks, some have supposed that the latter were not then domesticated among the Roman people: an opinion that would appear to have some foundation, as Pliny, about the same time, and Ælian, a century and a half later, both speak of that fowl as wild. But from the distinct mention made by Cicero (B. C. 60) of the manner in which ducks' eggs were hatched under hens, we are inclined to believe that the domestication of the DUCK may have been commenced (though in all probability very partially) before the period usually supposed.

Of all our domestic poultry, however, none appear to have been regarded with more importance, or to have been more esteemed, from the earliest ages, and among almost every nation, than those included in that cosmopolitan species, known *par excellence* as "COCKS AND HENS." The representations of them upon the coins, medals, and gems of the ancients, as also their dedication to Apollo, Mars, Minerva, Mercury, Jupiter, and other heathen deities and semi-gods, prove the reverential worship attached to them : and of all the myths of the ancients none strikes us as more beautiful than that by which the cock is rendered sacred to the first-named god, in the character of Phœbus, concealing as it does the fine simile of the dawn of day being ushered in by the voice of that " bird of morn," that doth

> " With his lofty and shrill-sounding throat,
> Awake the god of day ! "

The cock was also sacrificed to Æsculapius (the god of medicine) upon recovery from sickness. But it was not alone for sacrificial purposes that poultry were reared by the ancients; by whom their value and usefulness as articles for the table were fully appreciated. For that reason Pythagoras (B. C. 535) contends that fowls ought to have been cherished, and not sacrificed. So, also, Æsop ingeniously finds an excuse for the erratic gallantry of the cock, as being the means of providing eggs for the table of his master and mistress. Cicero, too, informs us that hens were reared and kept entirely for the profit derived from the sale of their eggs. According to Pliny, the inhabitants of the Isle of Delos were the first to fatten their fowls by artificial means (as cramming, &c.) ; and it was from them, he adds, that the rage for devouring fowls loaded with fat, spread like a contagion among the luxurious gourmands of his time, who spent their lives in endeavours to produce some unheard-of dishes and delicate *morceaux*. To Messalinus Cotta belongs the honour of first inventing the dressing of cocks' combs as a *fricassée*; whilst that more ferocious glutton Heliogabalus is said to have greedily devoured the raw crest just as it was cut from the living fowl.

At Rome those birds performed important services in enabling the augurs to read the oracles of Fate. The sacred pullets were kept in pens or coops; and, from their feeding, the professors of divination (hence called *Pullarius*) pretended to discern good or bad tokens. The mode of " working the oracle " appears to have been thus: at day-break the " Early Morning Augurs," (as they were also styled,) on being consulted upon any important event or undertaking, proceeded to scatter barley or other grain before the sacred chickens, and accordingly as they devoured the food greedily or otherwise, was the omen pronounced to be good or evil. Cicero exposes the absurdity of the auguries (*De Divinat.*) in a way that would seem to imply that those " Peep-o'-day Boys " sometimes withheld, on the night previous to an augury, the customary food of the chickens, in order to produce an evil omen !

The only other individual of our domestic fowls that we find noticed by the ancients, is the GUINEA FOWL, known by them as the *meleagris*, under which term it is mentioned by Ovid, Aristotle, Pliny, Varro, Columella, and other authors. But although we have abundant evidence of this fowl having been well known to the Greeks and Romans, (the latter esteeming it as a delicacy of the table,) we have no grounds for believing that it was domesticated among them; on the contrary, the allusions made to them lead us to suppose that the same wild, shy, and unsociable temper that even now, in some degree, prevents their more general and extensive domestication among us, operated as a bar to their complete naturalization in earlier times.

CHAPTER II.

COCK-FIGHTING AND OTHER ANCIENT CUSTOMS.—POPULAR
SPORTS, PASTIMES, &c., IN ENGLAND.

IF it has been said with truth that a volume might be written upon COCK-FIGHTING, assuredly another might fairly be filled with an account of the various other sports, pastimes, and customs, of which that fowl was formerly the subject: though we have no intention of perpetrating such an infliction upon our readers.

COCK-FIGHTING, it is said by some writers, was originally instituted by Themistocles (B. c. 480): others, however, ascribe the honour (questionable though it may be) to his contemporary, Miltiades. The story, as it is more generally accredited, being told by Ælian; who says that Themistocles, whilst leading his troops against the Persians, accidentally seeing a cock-fight, made use of the circumstance to endeavour to inspire his soldiers with courage and patriotic fire; and achieving a splendid victory over the enemy, the Grecians ever afterwards were accustomed to celebrate it by annual

cock-fights,—hence regarded in the light of con-
secrated and sacred institutions.

Plausible as may seem this story, we are rather
disposed to assign an earlier date to the introduction of that sport into Greece.   Palmerius, another
writer, mentions the institution of partridge and
quail fights as early as B. C. 550 ; and although
Mr. Pegge, in his learned paper upon the subject,
in one of the volumes of the *Archæologia*, disputes
this high antiquity for the sport, he does so, we
think, with but little show of reason.   Æsop, who
flourished about six centuries before the Christian
era, speaks in his Fables (in which he may be presumed to have portrayed the popular manners
and customs of the age in which he lived,) of a
partridge " turned among *fighting cocks* to feed."
Cock-fighting most likely originated long prior
to the time of Themistocles, among the nations of
the East, who, from the remotest period, appear to
have been (as indeed they continue at the present
day) passionately fond of these combats with fowls:
and in their probable introduction into Greece
from Persia, for the purpose of amusement, may
be discovered the origin of the appellation " *Persian birds* " applied to those fowls,—a name, perhaps, given to them, not so much to indicate their
own origin, (as some have supposed,) as in order
to designate the game or sport they gave rise to.

By whomsoever established among the Grecians,
we are certain that the Athenians were the first

(and probably the only) people by whom those fights were legalized and publicly enjoined and sanctioned. Cock-fighting speedily became a most favourite pastime; and the philosopher Chrysippus (B. C. 320) did not deem it unworthy of him to notice and extol the courage of the fighting-cock. Public shows or spectacles consisting of these combats took place annually among the inhabitants of Pergamos, where, according to Petronius, it was so prevalent that school-boys were promised a fighting-cock by way of reward. Upon a lamp found beneath the buried ruins of Herculaneum were represented two cocks in an attitude that denoted them to be engaged in a mortal conflict: and the representations of cock-fights repeatedly found upon the coins and gems of the Greeks, Dardani, and other people, evidence the extension of the sport, and national importance it acquired. It was subsequently practised at Rome; though it was not thought so much of there as at Athens,— the Romans exhibiting a far greater passion for partridge and quail fights, which hence became more general among them.

The introduction of cock-fighting into England has been generally ascribed to Julius Cæsar, but we believe erroneously; as the passage in his Commentaries quoted in support of the assertion, would rather appear to prove (conclusively to our view) the existence of that sport among the ancient inhabitants of Britain prior to the landing of the

Romans. Cæsar's words are,—" The interior part of Britain is inhabited by those who, by fame and report, are said to be born in the island itself. .... .. They do not deem it lawful to eat the hare, the fowl, or the goose : nevertheless these animals are cherished (or brought up) *for the sake of pleasure and diversion.*"

In this passage Cæsar doubtless refers to the existence of some semi-religious ordinance or prohibition of the flesh of those animals by the Druids; but it is quite evident that the primitive inhabitants of these isles were not only acquainted with fowls, but were accustomed to breed them for the sport and amusement they afforded.

The earliest distinct notice of cock-fighting in this country, however, is to be found in a very curious Description of London, by William Fitz-Stephen, a writer of the time of Henry II. In this work, which was first published by Stow in 1598, the author states that the pastime was so generally in vogue that it was the customary game even of school-boys. " Moreover, every year on the day which is called *Carnivale,* (as we have all been children, so therefore we begin with the games of children,) every one of the boys of the schools bring their fighting game-cocks to their masters; and that morning the whole leisure of the boys is given to the sport, to see the fight of their cocks in the school."

It is not surprising, therefore, that the pastime

thus encouraged (we might almost say, taught) in
the school-houses, should have grown into universal popularity, and have acquired the importance of a national sport. In the reign of Edward
III. so general had it become, that it was found
necessary, in 1366, to check it, in a proclamation
issued for the purpose, as it set forth, of putting
down the practice then so prevalent, of playing at
"Idle and Unlawful Pastimes," among which cock-
fighting was classed. Du Cange informs us that a
similar attempt was made a century before in France,
where combats with cocks were prohibited by an
Act of the Council. But it requires something
more than a royal proclamation or decree to up-
root or revolutionize the ancient sports or customs
of a nation; and cock-fighting continued to flourish
in this country: and in the time of Henry VIII.
even royalty itself had become reconciled to the
"Idle and Unlawful Pastime;" and a "Ryghte
royall" cock-pit was added to Whitehall by that
monarch, for his own especial amusement. In
consequence of the mischievous results attending
cock-fights, Queen Elizabeth, in 1569, issued a
royal proclamation for their suppression, though
it does not seem that her effort to put them down
was attended with any better success than the attempt made by Edward III., as Stow, the anti-
quary, writing at the close of the 16th century
says,—"Cocks of the Game are still cherished by
divers men for their pleasure; much money being

laid upon their heads when they fight in pits, whereof some are costly made for that purpose."

It is also recorded that even the rigid Scotchman, James I., indulged himself twice a week in the diversions of the cock-pit. Charles II. had one exclusively appropriated to himself and court, upon the spot now used for the Privy Council offices: and, it is said, that monarch was the first who instituted the sanguinary combat known as the " Battle Royal," (in honour of the regal founder,) in which a number of cocks, armed with spurs, were set to fight until one alone survived, to lament, as Alexander did, that he had no rival left to conquer!

Apropos of arming game-cocks with artificial spurs, Mr. Pegge says, that it was left to the refined cruelty of the English to invent that addition to the legs of the poor fowl in order to increase the bloody barbarity of the sport: but a passage in Aristophanes, which has since passed into a proverb, (*tolle calcar si pugnas*,) would seem to imply that this mode of shortening the duration of the conflicts was not unknown to the Greeks.

During the reign of the Merry Monarch, the sunshine of royal favour being extended to the sport, a passionate love for it pervaded every class of society, as we are informed by Burton in his *Anatomy of Melancholy*, wherein he gives a general view of the sports in vogue in his day. And in the present state of society, we are surprised to learn that ladies of rank and education used then to frequent and countenance the unfeminine sights of the pit: the pastime being apparently regarded rather as a polite and fashionable acquirement. Thus Powell in his *Cornish Comedy*, 1696, introduces a young heir of fortune, who with much point is made to exclaim,—" What is a gentleman without his recreations? Hawks, &c., and Cocks with their appurtenances, are true marks of a country gentleman! .... My Cocks are true Cocks of the Game: I make a match, and £100 or £200 are then soon won, for I never fight a battle under!"

In Scotland cock-fighting was formerly as usual at the schools as it was in Fitz-Stephen's time; and according to Brand, it continued in vogue until nearly the close of the last century, the masters encouraging the sport among their scholars, and claiming the craven cocks as their perquisite under the name of " fugees," or fugitives. It remained a favourite pastime in this country also, down to a period quite as recent. An attempt was made to suppress it in 1736, by a statute rendering it illegal to keep or provide public pits for such spectacles; but it was unsuccessful: and it was not until a sufficiently strong feeling against it was awakened in the public mind, that it was generally and effectually put down.

At the present day, we still occasionally hear of matches being clandestinely fought in some of our remote rural districts. In Spain, South

America, Mexico, and Peru, it is still extensively practised: in the latter state it forms the fashionable and almost daily amusement or occupation of the ladies. It is also general among the people of India, China, Malacca, Java, Sumatra, and of almost all the countries of the East, the inhabitants of which exhibit a most incredible and extraordinary passion for it: so great, indeed, is their love for the sport, that, as a recent writer assures us, in the Eastern peninsula of India, you can seldom meet a native without a fighting-cock under his arm; and as soon as he sees another with a similar pugnacious appendage, he sets his fowl down on the ground, (a challenge never refused,) and a combat there and then takes place forthwith.

Beside the sports of the Pit, there were a great many other cruel pastimes with cocks and hens very generally practised in this country about Shrovetide; the only amusement in which, for the most part, would appear to have been the barbarity and torture to which the poor fowls were subjected. It has been supposed that they nearly all owed their origin to the riotous revellings of the age in which the ceremonies and corruptions of the Popish carnival were instituted. At a very early period of the Christian church, the cock was selected as the emblem of watchfulness and vigilance. In this symbolical sense we find the clergy in the dark ages styling themselves " *Cocks of the Almighty*," as indicating their vigilant and watchful care in awakening the vulgar people to repentance, and in bringing them to the holy church. With a similar allusion also were vanes in the form of cocks (hence their name "weather-cock") erected on the top of church towers and steeples; the earliest instance of one so placed being recorded in the ninth century.

·One of the most popular of these cruel diversions was COCK-THROWING, — almost universally played at the annual wakes and fairs held at Shrovetide; although Shrove Tuesday was more especially devoted to that pastime. A live cock was tied to a pole or small stake, when as many as chose to play, amused themselves by casting short sticks, or cudgels, at the poor bird: this continued until some lucky fellow, with a more steady aim and savage thrust than the other competitors, managed to kill it by a fatal and (so to speak) humane blow,—the cock being claimed by him as a reward for his dexterity. In a MS. in the Harleian collection (A. D. 1344) there is a very singular representation of a school-boy triumph obtained apparently at this sport. Two boys are depicted carrying one of their companions who is holding a cock in his hands; whilst a fourth, in the rear, bears a banner on which is painted a cudgel, indicating the bird to have been won at cock-throwing. The extent to which this barbarous game at one time obtained among the people of this country, attracted the notice of our French neighbours, and subjected us to the severe censure and ani-madversion of their writers. Erasmus, writing in the commencement of the sixteenth century, sar-castically pretends to account for the custom, by saying that on Shrove Tuesday the English people devour immense quantities of pancakes, which drive them mad, when they immediately fall to the striking and killing all the fowls they see about them. But the inhuman practice was not confined to adults alone, school-boys varying the sports of the pit by it, as is evident from the writings of Sir Thomas More, who, describing the period of adolescence, speaks of his skill in " casting a Cok-Stele," or cock-stick.

Many and absurd are the tales that various learned authors have most ingeniously devised, in their attempt to explain (to their own satisfaction if not to that of their readers,) the origin of this cock-slaying game : these, however, we may un-hesitatingly dismiss at once, as purely speculative, and as nonsensical as they are imaginary,—readily admitting, at the same time, our own inability to furnish any better solution of the mystery, deeming it by far the wiser course (as another writer ob-serves) candidly to confess our ignorance, rather than to increase the accumulated mass of error by idle and useless conjectures. In all probability the origin of the baiting of birds is to be found in the same barbarous love of brutal amusement that gave rise to the baiting of bears, bulls, and other animals.

Brand, in his *Antiquities*, states, though without giving his authority, that cock-throwing was, in 1680, resorted to by the authorities of many of the metropolitan parishes as a means of providing funds *for the relief of the poor;* a plan worthy the attention of pauper farmers! Cruel as the sport was, it continued to disgrace our national pastimes, sanctioned by custom and unchecked by law, until the commencement of the present century. After it was abolished, wooden cocks upon loaded stands were substituted for live ones; but the charm of torture that attached to the " genuine articles " being no longer present, the mimic sport never was popular, and it became obsolete even before all traces of the diversion it was intended to supplant had entirely disappeared. The only relic of this once famous game, however, that has descended to us, exists in the cock of gilt gingerbread harmlessly impaled upon slender sticks, that make their appearance annually upon the cake stalls at the fast-disappearing country wakes and fairs.

Another obsolete custom was that of THRESHING THE FAT HEN, which Tusser, in his *Points of Husbandry*, 1580, styles one of the feasting days of the ploughman: but he alludes to it as if it were a pastime peculiar to Essex and Suffolk, although it was certainly practised in other counties as well. The manner in which it was played is thus described by Hilman, a native of Surrey, who, in

1710, published a rural calendar, (*Tusser Redivivus,*) and who had, it seems, witnessed the perform-.
ance of the ceremony in his own day :—

" The hen is hung at a fellow's back, who has
also some horse-bells about him ; the rest are
blinded and have boughs in their hands, with
which they chase this fellow and his hen about
some large court or small enclosure: the fellow
with his hen and bells shifting as well as he can,
they following the sound, and sometimes hit him
and his hen, other times, if he can get behind one
of them, they thresh one another well favour'dly.
. . . . . Afterwards the hen is boiled with bacon,
and store of pancakes and fritters are made."

In the Harleian MS. which we have already
had occasion to notice, there is a very curious
painting of a Joculator beating a drum or tabour
to a dancing dog, whilst a most ferocious looking
game-cock is likewise " tripping it lightly" upon
a pair of stilts. A facsimile is given by Strutt in
his *Sports and Pastimes*, but we are disposed to
think (as he anticipates his readers may) that it
was a mere trick of the illuminator's fancy, as we
have nothing to justify the belief that dancing
cocks ever formed part of our national amuse-
ments.

The Joculator above referred to, (the *Jugulour*
of the Normans, and the original of the Juggler of
later times,) was a person whose occupation consisted of " teaching animals to imitate the actions

of men,—to tumble, dance, and perform a variety of tricks." Anciently he was a most important personage, and filled an office in the royal household until the reign of Elizabeth, when falling into disrepute, he was no longer appointed.

The following account of another and even more ridiculous custom still, is extracted from the Fly-Leaves of the *London Librarian*, though upon what authority the writer gives it, we know not :—" A singular custom, of matchless absurdity, formerly existed in the English court. During Lent an ancient officer of the Crown, styled the KING'S COCK-CROWER, crowed the hours every night within the precincts of the palace. On the Ash Wednesday after the accession of the House of Hanover, as the Prince of Wales (afterwards George II.) sat down to supper, that official abruptly entered the apartment, and in a sound resembling the shrill pipe of a cock, crowed ' *past ten o'clock !* ' The astonished prince, at first conceiving it to be a premeditated insult, rose to resent the affront, but upon the nature of the ceremony being explained to him, he was satisfied. Since that period the silly custom has been discontinued."

At Michaelmas-time a custom existed among our fore-fathers of eating a green or stubble-fed goose, —which, it is said, owed its origin to a remarkable historical event. As the story goes, Queen Elizabeth, being on her route to Tilbury Fort, stopped by the way at the ancient seat of Sir N. Umfreville to partake of his knightly cheer and loyal hospitality: when, at the "Feast of Saynt Mychell," whilst enjoying a hearty dinner of "roste goose," a courier arrived with the joyful news of the loss of the Spanish Armada; and the queen, calling for a bumper of Burgundy wine to celebrate the event, its anniversary was ever after observed on that day, and is even now commemorated on our tables by a similar dish—the custom surviving whilst its origin is forgotten.

One other practice, as ingenious as it was cruel, at one time prevailed in many country places, of removing the soot from chimneys by tying a line to the legs of a goose, and then, letting it down and pulling it up again, it was compelled, by flapping its wings in its struggles, to supply the place of our modern chimney-sweeping machine.

CHAPTER III.

DOMESTICATION OF FOWLS IN ENGLAND.—IMPORTANCE OF
POULTRY SOCIALLY AND ECONOMICALLY CONSIDERED.

As we have already seen, the observation of
Julius Cæsar places it beyond a doubt that geese
and fowls were reared (if not domesticated) for
amusement among the primitive inhabitants of this
island, at least some eighteen hundred years ago:
and the earliest records that we possess, prove that
they did not long continue insensible of their value
for the purposes of food; as the Druidical injunction
against the flesh of those birds does not appear to
have been observed long after the Roman invasion.
Pliny enumerates among the choicest delicacies of
the early Britons, a kind of goose, which he calls
Chenerotes, and describes as smaller than the common goose of the Romans. This fowl has ever
formed a favourite dish in our country; and at the
present day it continues to enjoy unabated popularity.

The PEACOCK was probably introduced by the
Romans, the name bestowed on it by the Saxons

(*pawa*) evidently betraying its Latin origin (pavo). In the ages of chivalry valorous knights and courtly gallants paid their solemn vows before fair dames and the gorgeous birds of Juno. Formerly the pea-fowl does not seem to have been generally kept, being rather regarded as *rara avis ;* for it does not appear at any time to have been purchaseable in the markets, although swans and other birds of almost equal scarceness were to be obtained. It was, however, esteemed, and found a place at the sumptuous banquets of the great, rather as an ornament than a viand, on account of the beautiful appearance of its plumage, in which it was always served up. Of the manner in which it was cooked and made its *entrée* on the table, we have the following very singular account from the " *Form of Cury*," a curious roll of ancient *cuisine,* compiled by the master-cooks (the Soyers of their day) of Richard II. *circa* 1390; and which is quoted in the *Norfolk Archæology*, vol. ii. page 276.

" At a feast royall pecokkes shall be dight [prepared] on this manere : take and flee off the skynne with the fedurs, [feathers,] tayle, and nekke with the hed theron ; then take the skyn with all the fedurs, and lay hit on a table abrode, and strew theron grounden comyn ; then take the pecokke, and roste, and endore [baste] hym with rawe zolkes of egges ; and when he is rosted take him of, and let hym coole awhile, and take and sowe hym in his skyn, gilde his combe, and so serve hym forthe with the last cours. *Nota.* Pokok shal be par-boiled, larded, and rosted, and eaten with gynge-nere."!!

In 1430, at the coronation dinner of Henry VI., the second course comprised " pecocke in ha-kell." At the festivities given in honour of the marriage of the Duke of Burgundy and Queen Margaret of England, in 1468, it is said that no less than one hundred of those birds were daily served up for a week! Much later, too, we find peacocks, or *Peions,* occupying a conspicuous place in the lists of flesh for the table,—occurring fre-quently in the Household Books of Henry VIII. and his nobles. From Chaucer's allusion to " Pé-cocke Arròwes," it would appear that their fea-thers not only served to garnish a costly dish, but were sometimes used to fledge those warlike weapons.

Among the moderns the flesh of the pea-fowl is regarded (some writers assert undeservedly) as almost unfit for food; though Sir W. Jardine as-sures us that in India the native poulterers employ themselves in catching wild pea-chicks, for which they find a market at Calcutta.

Whenever SWANS were first kept, (we will not call their limited intercourse with society *domes-tication,*) either for "sight or service," their na-turalization in this country would seem to have been a labour both long and difficult in the accomplishment: and being of rare occurrence, they preserved, from the earliest period, a very excessive and extraordinary value in the market as compared with other kinds of poultry,—usually fetching the price of a quarter of wheat. They were always regarded as birds appertaining exclusively to royalty; and almost became objects of veneration among the people, being sometimes used as a pledge between the sovereign and his subject. Thus we are told, that in 1306, Edward I. swore a solemn oath to his nobles on two swans. Anciently these birds, when at large, could not be possessed legally by any subject, except by grant from the Crown: though there were doubtless many who, like Robin Hood, as the ballads inform us,—

" Swannes and Fesauntes had full good,
     And Foules of the Revere ! "

The legal title to hold them was, however, usually conferred by some distinctive device or *swan-mark*, which the party was specially entitled to use: but by a statute passed in the reign of Edward IV., this privilege was limited to persons seized of freeholds of the clear yearly value of five marks, (£3 6s. 8d.,) the king's sons being alone exempted from this restriction. In the time of Queen Elizabeth there were nearly 900 individuals and corporate bodies holding swan-marks, each having distinct devices, some of which were exceedingly curious, and are figured in Yarrell's *Birds*. The mark of the Company of Vintners (and which is used at the present day) was a notch or nick on each side of the swan's beak; a device still preserved (though perhaps unwittingly) by the many village hostellers, who glory in the sign of the "Swan with two necks," undoubtedly a corruption of the Swan with two *nicks*, or notches.

Anciently the Crown had swan-herds stationed on the Thames and in other places where Royal Swanneries were kept: besides these officers, there was a "Master of the Swans," whose duty it was to "supervise and approve" all swans being upon public streams or waters in the counties of Huntingdon, Lincoln, Cambridge, and Northampton. There was also formerly a court of "King's Maiestie's Justices of Sessions of Swannes," in which the regulation of the "Happing of Swannes" was vested.

Some of our readers may have heard of the Royal Swan Happing (or *upping*, as it was called by the Cockneys) annually in August; when the markers and other appointed conservators of the swans in the Thames, belonging to the Crown and the two city Companies of Dyers and Vintners, proceed up that river as far as Oxford (the ancient bounds) for the purpose of inspecting the swans, and of seeing them properly marked. According to the established regulations, upon these occasions all the young cygnets of the year, with any new-comers, are "*happed*," or snatched up into the boats, by those officers or their deputies, and marked sometimes with plain nicks upon the bill, but more generally with cross-bars formed diamond-wise. If .too many birds are found to have flocked to one particular spot, they also are *happed* and conveyed to other parts of the river. At the Swan Happing in 1841, the number of birds upon the river belonging to the Crown was found to be 232, whilst the Dyers' Company held 105, and the Vintners (who formerly possessed 500) only 100.

The corporation of Norwich still retain their swan-herd, whose duty it is to tend the swans upon the city streams. There is also a very singular custom still observed there, which is noticed by Mr. Yarrell, who derived his information from an interesting paper, communicated to him by the late Bishop Stanley, who then filled the president's chair in the Linnæan Society. Annually, upon the second Monday in August, the cygnets of that year, whether in charge of the city herdsmen, or belonging to individual members of the corporation and private persons, are collected together (to the number of some sixty or seventy) within one stream, where they are supplied with all the barley their rather voracious appetites can possibly dispose of, and in about three months they are considered sufficiently fattened for the table, when they are returned to their respective owners, accompanied by a copy of the following poetical directions for their treatment in the kitchen,—directions, by the way, so business-like as to merit a place in the next edition of Soyer's *Ménagere ;*—

TO ROAST A SWAN.

" Take three pounds of beef, beat fine in a mortar,
  Put it into the swan—that is, when you've caught her ;
  Some pepper, salt, mace, some nutmeg, an onion,
  Will heighten the flavour in gourmand's opinion.
  Then tie it up tight with a small piece of tape,
  That the gravy and other things may not escape.
  A meal paste, rather stiff, should be laid on the breast,
  And some whited brown paper should cover the rest.
  Fifteen minutes, at least, ere the swan you take down,
  Pull the paste off the bird, that the breast may get brown."

THE GRAVY.

" To a gravy of beef, good and strong I opine,
  You'll be right if you add half a pint of port wine ;
  Pour this through the swan, yes, quite through the belly,
  Then serve the whole up with some hot currant jelly."

" N. B. The swan must not be skinned."

Although in former times the swan was chiefly prized as a dainty dish " to set before a king," (so much so indeed that no state banquet was considered complete that did not include it,) its feathers, like the peacock's, were used in the manufacture of arrows for the wealthy. In an inventory of the effects of that martial knight, Sir John Fastolf, K. G., made in the middle of the fifteenth century, " viij. schefe arrowys of Swanne " occupy a place among other valuable articles enumerated.

DUCKS are no exception in regard to the obscurity which hangs alike over the primitive history of fowls and their domestication. The Rev. E. S. Dixon, in his amusing but rather speculative work on Poultry, considers they are an importation to our shores from the East Indies, through some channel or other; and conjectures that the approximate date may be fixed at about 1490, or shortly after the discovery of the passage of the Cape of Good Hope. This conjecture, we think, makes the introduction less remote then it really was. Ducks (and evidently tame ones) are mentioned in ancient records of a date at least two centuries anterior to that above named. Dugdale (*Antiq. Warwick.*) in a list of the prices of provisions in the year 1290, gives the value of a duck then at one penny: whilst by an act of the Common Council of London for regulating the sale of Poultry, the price of a mallard was fixed, about the same time, at three halfpence; thus showing the former to have been even more common than that truly old English bird. In the inventory of Fastolf's property, already referred to, mention is made of a cloth of arras showing some men shooting " a Doke in the watir with a crosse-bowe," and a " gentlewoman in grene taking a mallard in hir hondes." Chaucer too speaks of both ducks and drakes in a way denoting him well acquainted with their habits.

Anciently PIGEONS were more highly prized for the table than now; and their rearing was carried on extensively. Dovecotes were attached as a manorial right to the dwellings of the principal landed proprietors in England: many of these may still be seen, adding to the picturesque appearance of old country mansions. The number of these cotes scattered through the country, must have been very great, as under Charles I. they were made the subject of special enactment in two proclamations, (issued in 1625 and 1627,) "For the maintaining and increasing the salt-peter mines of England, for the necessary and important manufacture of gunpowder:" these decrees recite that, "whereas the realm naturally yields sufficient salt-peter without depending on foreign parts; wherefore, for the future, no Dovehouse" should be paved with stones, boards, bricks, lime, gravel, or sand; and an officer, styled a Salt-peter-man, was appointed, to go round and visit all Dove-cotes, houses, &c., in order to collect the nitric formations among the refuse of those places.

We have no means of fixing the date at which we made the acquisition in this country of the GUINEA-FOWL. Richardson states, on the authority of Kennett's *Parochial Antiquities*, that it was "well known in England as early as the year

1277:" but of this we have not the slightest confirmation, and, on the contrary, there is strong negative evidence on which to ground a belief that it was not introduced so early, nor indeed until after the Turkey, in the first part of the sixteenth century; as it does not occur in the sumptuous bill of fare of Archbishop Neville, in the reign of Edward IV.; nor is there any mention made of it in the Household Books (circa 1512) of the Duke of Northumberland, nor in those of King Henry VIII. If we were to hazard a conjecture as to the probable date of their arrival on these shores, we should say that it was not until after the discovery of Guinea by the Spaniards, in 1528, that the original of our stock were conveyed hither.

TURKEYS, according to the old and oft-quoted distich, were introduced here in 1524; and this, in all likelihood, is not far from being the correct period,—although Walcott, a writer of the last century, positively asserts that they were brought to England in 1521. The Rev. H. G. Dashwood states in the *Norfolk Archæology*, that the family of Sir George Strickland, Bart. have a tradition that one of their ancestors brought the first of that species of fowl to this country: perhaps, however, the family crest, *a Turkey Cock in his pride*, may have originated the supposition. In 1541, it ranked as a kitchen delicacy with cranes and swans; of which fowls Archbishop Cranmer, in his ordinances for the regulation of feasts, decreed there should only appear one dish each at any entertainment. Upon the occasion of a grand festival held in the Temple in 1555, turkey chicks made their appearance, and it seems they were purchased at 8s. the pair, being not much more than the ordinary price of fat capons, from which it is fair to infer that at that time they were tolerably plentiful. They speedily descended from the rank of a delicacy, and must have become the ordinary inhabitants of country poultry-yards, as old Tusser, writing about thirty years later, mentions "Turkeys well drest" among the dishes that were "counted good cheer" in the country at Christmas time. He also recommends the husbandman to

"Stick plenty of boughs among runcival pease,"

to protect them from peacocks and turkeys, which must evidently, therefore, have been very commonly kept in his day. The turkey did not reach France quite so early: the first intelligence we have of it there being in 1556, when a present, consisting of twelve of these fowls, was sent by the good people of Amiens to the king. In 1570 it formed a prominent dish at the nuptial feast of Charles IX.

In considering the importance of DOMESTIC POULTRY, we may conveniently do so under two heads :—

I. *Socially*, their uses and value as a means of

providing a healthful occupation and amusement for the people; and,

II. *Economically*, as an article constituting a considerable branch of trade and commerce.

On the first head little need be said: for there are few persons, no matter their sex or station, who have not at some period of their lives occupied themselves most agreeably in the rearing and tending some favourite occupant of the poultry-yard; and who do not remember the feelings of pleasure and gratification the caring for and watching these pet objects called forth at that time. There can be few, too, who would be naturally insensible to the easily accessible means of amusement that poultry-keeping (whether for the purpose of profit, or simply for fancy) furnishes to those who may seek to dispose of a portion of the leisure of every-day life in healthful and not altogether profitless recreation and enjoyment. And although the example of our gracious sovereign has, so to speak, elevated an otherwise humble and undignified pursuit into an almost aristocratic pastime, there is yet another point of view from which to hail the increasing and extended taste for poultry-rearing, so recently sprung up,—as being calculated to produce effects of the highest social importance: we mean the humanizing (we had nearly said, civilizing) influences it cannot fail to exercise upon the teeming masses of our industrious town-populations. This matter is thus ably touched upon by a recent writer :*—" It is some comfort to us at the present time to feel that we can see a bright side and a useful side to the Poultry Mania. . . . . We have a strong idea that the love of animals and the rearing of them are very humanizing and softening tastes ; and that if a mania suggests the care of them to the lower classes, (who, after their monotonous mechanical toils, want something *living* to take charge of,) it would be doing good service : those who live amid machinery, looms, shops, work-rooms, and factories, would be benefited by having their 'pets,' their domestic animals, at home—whether fish or fowl, dogs or rabbits. . . . . All these sorts of things do good—have gentle influences—keep the heart somewhat green in the midst of this dry, dusty world of ours, draw out feeling, and call forth a certain measure of affection.

" We are convinced that those who are engaged in the more sedentary trades need something growing or breathing — something of God's visible works—to keep them from depressing or self-centralizing thoughts, or from vacancy of mind altogether. The mere fact of having to take care of things, to feed or to water, somehow or other does good: and then there is a respond or return in the favourite animal, or the plant ; the flower breaks forth with grateful utterances as its leaves unfold before the master's eye, the dog licks its

* Nat. Miscellany, May, 1853.

master's hand, the bird leaps to the side of the
cage and puts its little beak through the bars, or
the fowls come scrabbling and skipping across the
yard : there is sympathy ; the workman's heart is
exercised—it is kept in play—it does not grow
quite hard or sour; the principle of living is preserved to some extent by such means as these ;
and if by any means we can but keep a man's heart
softened, and excite his sympathy in some direction or another, we may hope for the formation of
character, as there is ground to work upon, and
even high Christian principles may in time be
grafted in."

Although the rearing of poultry formerly constituted an important branch of the rural economy
of many nations, (as it continues to be at the present day on the continent,) still it is comparatively
but recently that public attention in this country
has been at all directed to the subject : whilst even
now it is very far from exciting such *general* interest as its importance demands and deserves.
Not that, speaking of its *importance*, we would
have it supposed that we employ the term in the
sense in which it is sometimes used by certain
poultry-maniacs, who seem to imagine that Cochins,
Spanish, and Dorkings, are about to supersede
Short-horns, Southdowns, and Leicesters ; for we
have too much faith in the beef-eating propensities
of Englishmen, and in the national prejudices and

predilections of our countrymen, to imagine that they will ever bring themselves to prefer the leg of a chicken to that of a prime wether; or to assent to the substitution of the frivolities of fricasées and fritters for the substantial realities of rump-steaks.

The art of poultry-keeping seems never to have been so well understood, nor its importance so generally appreciated, in England as in France. In the latter country, in many of the departments, the rearing of fowls forms the principal rural pursuit of the inhabitants, and of sufficient importance and profitableness to encourage the farmers to raise a very considerable quantity yearly of maize and buck-wheat, merely as food for the poultry. In the pursuit a much larger proportion of the agricultural population is engaged than in our own country. There a *fermier* of only a middling occupation will most usually have as many as 500 or 600 fowls in his poultry-yard: and yet poultry is much dearer in England than on the continent, as must be evident from the large quantities we are supplied with from abroad. The *Morning Chronicle* commissioner, who visited France to report upon the state of its agriculture, was struck with the importance that poultry-rearing assumed there, constituting as it did, in many parts, the material element of agricultural prosperity. He was astonished to observe " the immense quantities of fowls, ducks, geese, and turkeys, reared principally for the purpose of exporting them or their produce to the London markets : for miles and miles the country was alive with them."

From official returns it appears that, during the last two years, the declared value of live and dead poultry imported from abroad and entered for home consumption has been as under,

1851 . . £31,523 | 1852 . . £34,130,

whilst in the first five months of the present year it has been £6,780, being little more than half the average supply in a similar period of former years. The principal portion of the above supplies has been derived from France and Belgium ; with, recently, some additions from America. In very severe seasons large importations of wild fowls are also received from Sweden and Norway,—as was the case last spring, when an unusually late consignment of 25,000 head of game reached the London markets : it comprised ptarmigans or white grouse, capercaillie or cocks of the wood, and some black grouse.

As an illustration of the careless indifference with which the commercial view of the subject has been hitherto regarded, we may remark that about £200,000 are annually paid by our countrymen, for the items of imports embracing poultry and its produce, needlessly to strangers ; all of which might very much better be spent among themselves, and put in their own pockets—a matter very easily effected, " provided always " our farmers were not too high-minded duly to regard such small affairs as the rearing and feeding of chickens and fowls.

In addition to our foreign supplies, poultry is largely imported from Ireland, in some parts of which country they are very extensively kept. According to agricultural returns and estimates there were throughout the provinces the following numbers of poultry :—

In 1847   .   .   .   5,691,000 head.
1849   .   .   .   6,326,000 —
1851   .   .   .   7,013,000 —

As much as ten tons of poultry and fifty tons of eggs have been shipped from Dublin for Liverpool in a single day.   The fowls exported are not, however, of a superior description, as they are said to partake too much of the proverbial *greenness* of the Isle.

For the metropolitan markets the greatest home supply of chickens and fowls, and of the best quality, are received from Surrey and Essex: the same may be said of geese from Lincolnshire and Suffolk; of ducks from Buckinghamshire; of turkeys from Norfolk and Cambridgeshire; and of guinea and pea-fowls from Essex alone.

Wokingham, in Berkshire, is also particularly famous for fowls, (especially fattened,) which constitute the chief trade of the town; and by it many families there and in the neighbourhood gain a livelihood. The fowls are sold to London dealers, from whom as much as £150 in one day have been returned in this traffic: and for one gala at Windsor there have not unfrequently been purchased formerly as many as 20 dozen, and that too when they realized half-a-guinea a couple. Romford, in Essex, is also a great market for poultry, but that generally of the store or barn-door kind, and not artificially fed or fattened. Fowl as well as goose feeding is carried on to a far greater extent in the vicinity of London than in most other parts of the kingdom. At Stratford and Bow the fattening of fowls is as systematically and successfully conducted as that of geese; and it is said that the despatch with which the feeding is accomplished far exceeds anything witnessed elsewhere. From Stratford, too, large numbers of fatted guinea-fowls are yearly sent to London by the Messrs. Boyce, the experienced poulterers.

In some agricultural districts geese form an important branch of traffic. In Wokingham, and the neighbourhood thereabout, they are bred in large numbers, and about Midsummer they are sold to itinerant dealers, who in turn dispose of them to the poulterers in the different localities near London; where, it is not so generally known, goose-feeding is carried on to such an extent that one person feeds upwards of five thousand every season. For the best geese, however, we are undoubtedly indebted to the Eastern and Fen counties of Norfolk, Suffolk, and Lincoln : in the latter county they are regarded as the staple produce; and the reader may form some idea of the scale on which goose-breeding is conducted there, when we state that establishments are not unfrequently to be seen in which upwards of five coombs of corn are daily consumed by the brood geese alone : whilst the feeding is even more extensively carried on,—for instance, in the year 1850, no less than 100 quarters of grain were consumed every week during the season, at one farmstead, at Monk's Dyke in the same district, on which 3210 geese were fed daily. Several thousands of these fowls are annually fattened and sent to the metropolis by the celebrated poulterer Mr. Bagshawe, of Norwich ; upon whose extensive grounds above 2000 geese (besides other birds) are sometimes fed at one time : this supply of stock is not exclusively obtained from the county, a large number being received from Prussia, Belgium, and Holland.

Ducks, in vast quantities, are reared and fed in Buckinghamshire for the London markets, at which they maintain their character, and are much in demand. In that county a large portion of the poorer population obtain a living entirely by the rearing of that class of fowls. In addition to the places already mentioned as chiefly furnishing the supplies to the London markets, we may also name the counties of Herts, Devon, Somerset, and Wor-

cester, as producing poultry in abundance and of
fine quality.

From the statements given above, it must be
evident that, in those districts distinguished for
poultry-breeding, the greatest numbers are reared
and fattened for the empire city of the world;
some statistical details, therefore, of the supplies
furnished to the metropolitan markets may not be
unacceptable, forming as it does a branch of business in which many farmers are concerned to a
considerable and profitable extent.   It is only
right, however, to state that for the following particulars we are mainly indebted to the valuable
and original work of Mr. H. Mayhew, entitled
London Labour, and the London Poor.

At the commencement of the present century,
the country people in the outskirts of London used
to be the chief purveyors of poultry and game to
the city: they used to buy up the loose stocks of
fowls, and those being conveyed to town in waggons were hawked about from place to place.
More recently geese and turkeys were driven up
to London in a store (or live) state, in flocks of
from three to six hundred, and were disposed of
at a low price for the purpose of fattening as
Christmas fare for the suburban residents: the
fowls had thus often to travel more than a hundred
miles in a limited number of days; but from the
first improvement in our mode of road convey-

ances, their passage to London was generally made by land carriage, some few flocks still occasionally travelling as before. Very considerably smaller live stocks are even now sometimes disposed of in a similar manner, and for a similar purpose; but the introduction of railways, and consequent facility of communication, affording a rapid transit by which dead poultry can be brought up in a fresh state, the demand for live fowls has very much diminished, and immense numbers are now fattened and killed in the country, being sent up dead in hampers: thus has the trade of the metropolitan feeders fallen off in a ratio corresponding with the increase in that of the provincial dealers. Live fowls are also in request among the inhabitants of the suburban districts, as a means of providing themselves with the luxury of new-laid eggs: the best fowls for stock are obtained from Surrey and Sussex; and an inferior class (which forms three-fourths of the whole traffic) is brought from Ireland. These fowls are vended through the streets at cheap rates by hawkers. These men used also to deal in game received from poachers, which they used to sell to hotel-keepers and wealthy traders; but the legalizing the sale of game, in 1831, not only put an end to this illicit trade, but almost annihilated the street sale altogether. At the present time, Mr. Mayhew computes the number of Hawkers of Poultry and Game to be between two and three hundred. The bulk of the turkeys that make their appearance in London are those fed for Christmas, or in the months immediately preceding and subsequent, when the quantities (fattened) sent up from Norfolk alone are incredibly great.

In December, 1793, the number of turkeys sent up by the stage-coaches from Norwich, and consigned to the poultry-dealers in the metropolis, amounted to 2500, and weighed nearly fourteen tons. The week preceding Michaelmas, 1830, no less than forty tons of poultry were sent in waggons from Bury St. Edmunds, in Suffolk, to London; and of this quantity three-fourths were geese, sixteen tons of which were the property of Messrs. Flat and Walton, poulterers, of Tostock and Hepworth in that county. Mr. Clarke, of Boston, (Lincolnshire,) at Christmas, 1833, transmitted the following number of fowls to London,—geese 2400, turkeys 800: whilst Mr. Haines of Spalding killed for Leadenhall market, 1150 geese, 500 Turkeys, 360 fowls, and 200 ducks. During the same season also upwards of twelve tons of poultry have been imported in one week at Dover from France,—also destined for the supply of the metropolis. In 1848, during one week in September, above thirty-seven tons of poultry were conveyed from Norfolk by the Eastern Counties Railway. And in addition to these supplies, an immense number of fowls are sent up by private friends as Christmas presents, from the Norwich, Lynn, and Yarmouth markets, as well as other places in Norfolk, Suffolk, and Cambridge. These supplies, there can be no doubt, so far from falling off, are annually increasing greatly.

The poultry markets of London are Newgate and Leadenhall,—the latter being the largest in the world, and about two-thirds of the whole number of fowls sold in London are vended there. According to Mr. Mayhew, in his work already referred to, the total number of fowls sold annually in the above markets is estimated as follows :—

POULTRY SOLD AT NEWGATE AND LEADENHALL MARKETS.

	ALIVE.	DEAD.	TOTAL.
Game, &c. . .	. . .	910,000	910,000
Domestic Fowls	60,000	1,756,000	1,816,000
Geese	. . .	1,002,000	1,002,000
Ducks . . .	40,000	383,000	423,000
Turkeys . . .	. . .	124,000	124,000
Pigeons . . .	. . .	383,000	383,000
Totals	100,000	4,558,000	4,658,000.

To the above may be added about 350,000 heads sold by the poulterers and shop-keepers, making a total of about five millions of birds, wild and domestic.

Eggs also constitute a by no means unimportant article of commerce. The largest supplies to the English markets are obtained from Ireland: from which country the annual exports of eggs, according to official returns, amounted in 1835 to seventy-two millions; but now it is estimated that we yearly receive one hundred and fifty millions. Of this number London and Liverpool respectively consume twenty-five millions each.

The chief imports from the continent are from France, some from Holland and Belgium, and occasionally from Spain and Portugal. Half of this supply is taken off by the consumption of the metropolis. Besides being valuable as an article of food, eggs are largely employed in the manufacture or dressing of kid gloves, the yolks being used to soften the leather: in this way upwards of 80,000 eggs are annually consumed in one large glove factory at Bermondsey; the eggs when imported being placed in lime water, by which they are kept good for use a whole twelvemonths. From the returns of the Board of Trade, it appears that the *average* number of eggs *annually* imported and entered for home consumption in Great Britain, during the period ending 1828 to 1832, was 61,431,062; from 1833 to 1837, it was 68,493,516 ; and from 1838 to 1842, the average reached 91,393,732 : since which the yearly imports have been as follow :—

1843 70,415,931	1848 88,091,277
1844 67,565,167	1849 97,884,557
1845 75,627,362	1850 . . . 105,780,540
1846 72,299,487	1851 . . . 115,524,243
1847 77,542,311	1852 . . . 108,320,490

Whilst in the five months ending May 1853, the number entered for consumption has amounted to 51,694,026, showing a diminution in comparison with former years. The weight of the annual imports is about 4500 tons; and the value of the same reckoned at only 4d. per dozen eggs will be £150,000, which would be very considerably enhanced if we added the extra charges for freight, duty, &c.

More insignificant, though scarcely less necessary, articles of commerce in connexion with poultry are the feathers and down. From the earliest times they have been eagerly collected and sought after to furnish the luxury of soft couches and beds: and under the reign of Henry VI. their value caused them to become the subject of a statute or enactment. At the present day the down and feathers of ducks, pigeons, and partridges are used in France for pillows and mattresses. In the Lincolnshire fens, where goose-rearing is very extensively carried on, the feathers are considered most valuable; as for stuffing beds those of geese are esteemed more suitable than others. Whether from increasing desire for luxuries, or the dimi-

nution in supply, or from both causes co-oper-
ating, the demand is obliged to be supplied by importation; and the value has consequently been
maintained, and indeed increased. The down of
the Eider duck is collected in very considerable
quantities by the inhabitants on the coast of Sweden
and Iceland, where the fowls locate themselves in
great numbers. The Eider down (of which each
duck yields about one pound) is rendered so valuable by its lightness and elasticity, which is so
superior, that two or three pounds, whilst capable
of being compressed into a ball a man might hold
in his hand, are sufficient to fill a foot-mattress for
a bed. The declared value of foreign feathers imported into this country

In 1846 was . . £5279	In 1848 was . . £4680
1847 . . . . £4237	1849 . . . . £5096

Messrs. Hering of London have recently introduced, and successfully, the purification of feathers
on an extensive scale by steam machinery.

CHAPTER IV.

NATURAL HISTORY OF GALLINACEOUS BIRDS. CLASSIFICATION, NOMENCLATURE, AND COLOURS OF FOWLS. SPECIES AND VARIETIES.

UNDER the term *Domestic Poultry* are generally comprehended, in this country at least, the common fowl, turkey, guinea-fowl, goose, and duck; to which may now perhaps be added the pigeon, although this latter, together with the swan and pea-fowl, would be more correctly described as ornamental poultry. The wild breeds of those species are generally known as *Game*, and include pheasants, partridges, grouse, water-fowl, &c.: which are objects of pursuit for pleasure to the sportsman, and also (in the case of water-fowl more especially) to those residing on the sea-coast, and in the vicinity of lakes and rivers, where they are taken in decoys for market. In ornithology, however, the heterogeneous members of the poultry-yard are separated, and distributed and collected under their respective orders and *genera*, or families. Thus the gallinaceous fowls are comprised in the order of birds called *Rasores*, or scratchers; and which also include, according to the systems of modern writers, the family of *Columbidæ*, or pigeons; the web-footed or duck tribe being classed in another order, called *Natatores*, or swimmers.

The order Rasores is divided into five great genera or families; but it is only with two of them that we have anything to do,—namely, *Pavonidæ* and *Columbidæ*. These families are again divided into sub-families or smaller groups. Poultry are comprehended under the following classes, in the arrangement of which it may be remarked that we have alone consulted the form most convenient for the design of our work, and have not adopted it with any view to scientific accuracy.

Ord. RASORES; *Gen.* PAVONIDÆ.

Sub-fam.	Type.	Sub-fam.	Type.
Gallus,	The Cock.	Meleagris,	Turkey.
Phasianus,	Pheasant.	Numida,	Guinea-fowl.
Pavo,	Peacock.		

Gen. COLUMBIDÆ.

Sub-family, Columba,—*Type*, Pigeon.

Ord. NATATORES; *Gen.* ANATIDÆ.

Cygnus,	The Swan.	Anas,	The Duck.
Anser,	The Goose.		

The general characteristic points of the Rasorial birds are these: body large, heavy, but compact in form, with full, ample chest; the general plumage loose, and of every possible variety and brilliancy of colouring; wings mostly very short, (compared with size of body,) rounded at the sides, and hollow beneath, with the quill or flight feathers weak, and the bone upholding the shoulder only imperfectly developed, thus rendering the organs of flight weak, and the act itself both laborious and difficult, if not almost impossible; the limbs at once large, muscular, and powerful, the foot-bones of the legs (which are covered with strong, tough, leathery scales) terminating with three toes in front on each; immediately above and at the back of these a fourth or hind toe springs, this latter being both by its length and its position shorter than the fore toes, only just resting on the ground in the action of walking; these toes are bound or united together at the bases by a sort of membraneous ligature; claws curved downwards and strong, well adapted for scratching on the ground in search of food, &c.; the legs of the male birds usually furnished with strong, horny "spurs," employed as weapons for fighting; head and neck small (rather) and symmetrical; the former, in a greater or less degree, adorned with a crown or crest consisting of an upright, naked, fleshy excrescence; patches of a similar substance extending backwards over the cheeks to the ear-lobe, with wattles of the same drooping from the under bill beneath the throat; bill short and stout, the upper one more or less curved on the outer ridge; tail consisting of from ten to eighteen feathers, in the male often highly coloured, and so large as to encumber the bird and add to the great difficulty of flight. The food consists of grain of all kinds, roots, insects, &c., which the birds of this order eat indiscriminately; they are polygamous in their intercourse, and very prolific, producing numerous eggs, and hatching large broods of young at a time; these latter make their entrance into the world partially fledged or covered with down, and are at once able to run alone.

The family of *Columbidæ*, now generally included in or attached to the Rasorial order, must be excepted in many important particulars from even the *general* characteristic points enumerated above: many of the birds of this family, for example, (as the carrier pigeon,) possess very remarkable powers of flight, are monogamous, attaching themselves, if they are allowed, to one solitary mate through life, lay but two eggs at a time, and the young are hatched blind and perfectly unfledged, requiring the most anxious care of the parent birds, who are also compelled to feed their young for some time after hatching,—thus presenting some very striking points of difference between their general economy and that of the other members of the Rasores.

The Natatorial or aquatic genus comprehends many subdivisions, of which it will only be necessary to describe the general characteristics of the Anatidæ or duck tribe. The body large and boat-shaped, well adapted for floating upon water; the plumage or feathers are of close texture and possess a sort of water-proof or repellent property, casting water off as soon as it touches it; the wings of moderate size, the first quill of nearly the same length as the second; but, with few exceptions, little or no powers of flight; the legs are short and strong, but so set or drawn backwards to the belly, as to give the bird an exceedingly awkward gait, familiar to all as a "waddle," and giving the fowl the appearance of being almost overbalanced by the fore or anterior portion of the body not receiving support; the legs terminating with three fore toes on each, which are wholly united together by a thick membraneous web; a fourth or hind toe detached from the others, and joint higher up the shank of the foot bone; neck rather long, and arched, or inclined backwards; the bill large, straight, and powerful, generally depressed in a greater or less degree towards the point, which is round and blunt, and sometimes very broad: the bill is covered with a tough skin or membrane furnished with sensitive nerves; at the base of the upper one are the nostrils, appearing like oval lateral slits, half closed by the membraneous covering; the edges of the bills are supplied internally with rows of tooth-like *laminæ*, serving as a sort of comb or strainer, through which the bird is enabled to drain off the water and mud from the pools it inhabits, retaining the worms, insects, and vegetable substances: like the gallinaceous fowls, they produce very many eggs, sometimes, it is said, more than they can hatch at one sitting; like that order also, the young are hatched partially fledged, or covered with down: they devour greedily almost any description of garbage, or animal and vegetable matter, also slugs, aquatic insects, &c. Though some live almost entirely in the water, and others partially on land, some, like the goose, scarcely ever resort to the watery element unless as a refuge from anticipated danger. Nevertheless, the old English proverbial query of "Can a duck swim?" sufficiently expresses their general habits and love for water.

Having thus briefly sought to lay before our readers a sketch of the natural history of our domestic poultry, divested of any assumed scientific diction, and sufficiently popular to be practically useful in rendering perfectly intelligible the descriptions of the various wild and domestic species, which follow in course,—we proceed next to offer some observations upon the classification and nomenclature of the inhabitants of the poultry-yard, with a few suggestive remarks as to the colours of their plumage and the mode of describing them.

It is a remarkable evidence of the undeserved neglect in which our domestic fowls have been suffered so long to remain, that no systematic arrangement or classification of the species has yet been determined on, nor indeed even attempted.

It is true that various artificial plans have been proposed, but these have rather been urged with the apparent object of carrying out some particular crotchets of the originators, than with the view of satisfactorily determining the order in which they should be classed; thus we have had fowls assigned places either according to the size of their combs and wattles or the length of their tails, whilst even their legs and spurs have been also by turns pitched upon as the distinguishing marks by which to separate a group of fowls: to most of which, however, the objections must be obvious to all who are acquainted with the subject. It is probable that no *accurate* classification can be devised at present, with our scanty knowledge of the wild species of our poultry, (more especially of the genus *gallus*,) among which we believe the types or originals of most of our distinct domestic breeds might be discovered.

In the mean while the order we have pursued in describing the different species may be found at least as useful and easy of reference as any of the random classifications (if they can indeed be termed so) hitherto adopted by writers on the subject; whilst at the same time it will throw no difficulty in the way of the reader in tracing out any supposed resemblance, or in marking the distinctive features and characteristics, of the numerous species and varieties into which the poultry family are split up.

If the arranging or classifying of poultry is a
matter of difficulty, the proper and correct *naming*
of them is scarcely less so.   Every one connected
in any way with poultry must at some time or
other have experienced the inconvenience and serious annoyance resulting from the indiscriminate,
not to say reckless, manner in which individual
fowls, of almost every class and family, are chris-
tened with a multitudinous array of names of
the greatest diversity and dissimilarity, at the
will and caprice of imaginative individuals,— as
also with the local or provincial nick-names by
which they may happen to be familiarly known
among the country boors who are to be found in
most of our rural districts. " Nothing " (writes Mr.
Baily, the eminent poulterer) " is more tiresome
than the labyrinth into which the fancy naming of
fowls involves the inquirer, [and practical man
too,] and nothing is more vexatious than to find,
after having, with much trouble and labour, sought
to see the *new* breed, that it is only an old friend
under a new name : but such is often the case, and
the deception is innocently caused by the owners of
the birds."   This confusion and these errors arise
in many ways : if a poultry fancier meets with (to
him) an unknown or uncommon fowl, he immediately sets about naming it, and racks his wits to
discover the name of some outlandish person or
place, with the view of demonstrating to his neighbours the undoubted novelty and distinctiveness

of the breed, and the remoteness of its origin and source: thus when the black Indian ducks were first imported into this country, every lucky possessor of the importations was resolutely determined to engross the exclusive copyright (if we may so phrase it) of the original and genuine article; and therefore in the exercise of his "fancy free," he gave them the name calculated the most effectually to conceal the real quarter whence the stock was obtained: and as an illustration of the consequences hence resulting, we may enumerate the following as *some* of the names by which this valuable species is now known,—or rather, under which it is now disguised;

East Indian,	Madagascar,	Buenos Ayres,
West Indian,	Botany Bay,	Blackamoors,
Labrador,	Hudson Bay,	Beavers.

Thus it appears that they already almost rival the ancestral Welshman who gloried in the endless patronymics of Ap-Davids, Ap-Evans, Ap-Owens, &c. In a similar way, if any new variety (or more properly, variation) is produced in a particular species either by accident or hybridising, the most ambiguous, *mal-à-propos*, and deceptive name possible is given to it :—thus we have the Pheasant-Malay, probably a cross between the Game and Hamburg fowls, but certainly with no more connexion with the pheasant than with the sea-serpent;

so also there are Owl-pigeons with about as much of the owl as the ostrich in them.

No one can have failed to remark that poultry-dealers never know the pedigree of the fowls they have for sale, of new and uncommon breeds, nor where they come from; unless, indeed, as one is sometimes assured, they are from the antipodes; though, more generally, they were " picked up out of the way," or came in with another lot.

It is equally inconvenient and more to be regretted, when parties learned in the poultry-fancy, take on themselves, for no apparent advantage to any one, to alter the names that have been originally applied to certain species, and by which they have been previously designated and generally known: and it is therefore to be hoped that the gratuitous and uncalled-for change, recently proposed by some writers in the *Cottage Gardener*, in the name of the Cochin China fowls, by the substitution of " Shanghai," will be most studiously avoided and resisted; in order that the confusion that must arise (and which its very partial adoption has even already caused) from such a change may at the outset be prevented.

With the view therefore of removing, if possible, the difficulty that might be felt in attempting to identify any fowls under their various designations in the succeeding descriptions, we have adopted the plan of placing them under the several names by which they are most generally known in this country, adding such others—whether vulgar and popular or otherwise—as may serve to enable even the most uninitiated reader to discover the variety he seeks under whatever name or dress it may be enshrouded.

Before entering upon the descriptions of the fowls, the erroneous ideas that exist respecting their *colours*, and the totally misapplied and incorrect terms by which the varied hues and rainbow tints of the plumage of poultry is too frequently sought to be depicted and conveyed, seem to render some explanatory remarks on that head necessary. Much of the confusion referred to now prevailing has been introduced along with the Cochin Chinas; the ever-varying colours of which have not been sufficiently attended to; they have been loosely and carelessly described, no attention being paid to the employment of appropriate words in which to delineate them, and as a necessary consequence, the most absurd ideas are conveyed; the terms now in vogue giving only a faint idea (and often quite an opposite one) of the true colours or shades. We have frequently heard shades of light buff described as " lemon " or " silver pheasant," dark buff and brown as " ginger " or " partridge ; " a deep red or chesnut as " cinnamon," with many other absurd and ridiculous phrases, that convey just about as correct a notion of the true colours, as would the very intelligible term " ginger-blue " of the niggers, in whose dialect it is said to indicate *yellow and scarlet*. It is probable that to the use of similarly incorrect terms in describing the game fowl, we owe many of its numerous varieties, which some persons really believe in, but in whose existence or identity we have no more faith than we have in that of " Mrs. Harris."

The colours and even shades of colours of a fowl should be most carefully studied and compared before any attempt at describing them is made : equally important is it also that in doing so there should be employed the names of the most *common* and *well-known* colours only ; or comparisons with the shades of the most *familiar* objects of every-day life : the use of the names of *other animals* or *fowls* (which possibly the party meeting with the description may never have seen any more than the fowl described) ought never to be used,—unless indeed *every other mode of expression fails,* when of course the nearest approximate idea must be adopted : but even in that case the name of the fowl selected as a comparative should not be prefixed or conjoined to that of the fowl intended to be described,—as for example, " Partridge Cochin," " Silver Pheasant Cochin," &c. ; by which phrases a confused notion of some imaginary connexion or affinity often in course of time obtains : and surely it would be much better (even supposing the comparison necessary) to say " Partridge *coloured* Cochin," " Silver Pheasant *coloured* Cochin," &c. ; for although it may be urged that those names would not be so smooth or fluent, that objection cannot be allowed to weigh against the very serious inconvenience and most mischievous confusion induced by the use of the other terms.

In accordance with the views above expressed, we have endeavoured in the succeeding descriptions to employ only the most plain, familiar, and unmistakeable terms in depicting in language the colours of the fowls; bearing in mind, as we conceive all writers should, that such descriptions are needed not by readers to whom the birds are well known, but by those who seek to become acquainted with them through the medium of books. As, however, the want felt for suitable and familiar words in our language has in some instances compelled us to resort to the use of those of less frequent occurrence, or less generally understood meanings, (and which we would gladly have avoided if it had been possible,) we subjoin such, with brief explanations of what we intended to convey thereby.

Black,—a dead, full, and almost palpable black colour.

Lustrous Black,—a rich, shining, transparent kind of black.

Blue Black,— the two colours blended and glancing into each other as it were ; the outlines of the two undefined and inseparable.

Steel Blue,—the deep, bright, nearly purple blue assumed by steel springs.

Metallic,—the indescribable rainbow hues and tints seen on live fish, on some minerals and ores, and on bright steel when placed in the fire.

Golden Red,—the rich radiant hues of the setting sun : the *deeper* shade, so described, nearly an orange.

Golden Green,—the two colours blended together as in the plumage of some parrots.

Maroon,—a rich, velvet-like, deep brown puce.

Dickson and some other writers assert that "the colours of fowls frequently change in a very surprising manner;" and that hens "assume the plumage of the cock bird." This is true, except that, instead of such eccentricities of nature being of frequent occurrence, they are not at all common : true it is that we are sometimes puzzled, and even find it impossible, to determine the precise colour a recently hatched and only partially feathered chick will ultimately be, and that as it puts forth its plumage gradually, colours become apparent that were not before observed, and of which the bird gave no previous indication ; but the instances recorded of an absolute and *bonâ fide* change in colour, and of the assumption of male-plumage by the hen, are so few, as to prove them to be purely exceptional ; and in the latter case it is probably the result either of sterility, disease, old age, or a cross in the breed.

WILD SPECIES.

Sub-genus Gallus.

I. GIGANTIC FOWL.

Name.—Chittagong; Kulm, Dukhun or Deccan, Herat, and Jago Fowl. Gallus Gigantéus of Temminck.

Description.—Of large size, standing from 27 to 30 inches from the ground to top of comb; and measuring from the tip of the bill to the end of the rump 24 inches: the general plumage brown, or deep golden red; the neck, neck and rump hackles, (all very long,) a lighter shade; the back and lesser wing coverts a deep brown or chestnut, and the feather webs separated: wing coverts and tail (the latter large) entirely of a rich dark green: the under portion of the body of a metallic green and black: neck long, throat partially bare; head and wattles small, as compared with general dimensions; comb commencing at base of bill and extending some way back over the head, but short and thick, presenting the appearance of having been torn off: legs and thighs very long and thick.—The hen about half the height and two-thirds the size of the male bird.

History.—This fowl was first seen and described by Dampier and Marsden in their voyages and travels: and from these sources it has been subse-

quently more fully described by Temminck.   It is
a native of Java and Sumatra, whence it has been
introduced by the Mussulmen into India and Hin-
dostan.   Though wild it has been domesticated
in the latter country, and is frequently met with
in that state in the province of Deccan and Hyder-
abad, on the Kistnah river.   Col. Sykes, in June
1831, imported two cocks and a hen into England,
the winter climate of which they bore well.   The
hen laid freely and reared two broods within about
a twelvemonth after landing.   In 1835, Keith E.
Abbott, Esq. presented a cockerel and two pullets
to the Zoological Society : these came from Herat
(in Khorassaun), after which place he named
them.   Sir W. Jardine states that a specimen from
Sumatra is to be found in the Edinburgh Museum.
More recently others have been brought to England.

II. BRONZED FOWL.

Name.—Coq bronzé ; Gallus Æneus of Temminck and Cuvier.

Description.—Smaller (much) in size than the
preceding : general plumage of a rich ruddy purple and green, metallic, or, as its name implies,
bronzed : the feathers of the neck long, but not
running into a hackle, and of a metallic green ;
wing coverts ruddy purple ; tail, very full, of a
metallic purple and green, with rump hackles of

the same; throat, breast, and under portions of the body a bright lustrous black with shades of bronze; legs strong, of a light bluish ash colour; comb large, whole and not indented on the ridge, scarlet; wattle thick and small; face and throat bare, but red. Hen not described.

History.—It is a native of Sumatra, in the interior of which country it abounds; but only one specimen has reached Europe, that being sent to Temminck by M. Diard.

III. FORK-TAILED FOWL.

Name.—Gallus Furcatus of Temminck; Gallus Javonicus of Horsfield.

Description.—In size rather smaller than the gigantic fowl; the length from the extremity of bill to the extremity of tail being two feet: general plumage, steel blue, green, and golden hue, running into or blending with each other as the plumage of a green parrot or peacock; neck and chief part of the back being covered only with short rounded feathers; rump hackles, long and drooping, black shaded or marginated with pale yellow; wing coverts black shaded with deep orange; tail fork-shaped, black, with two or three metallic green scimitar-like feathers; under portions of the body shining black; the legs rather long, and yellow; the face bare, but bright red; comb of same colour, not large, single and whole; the bill yellow, beneath a large single wattle depending. The hen is very inferior in brilliancy of plumage : head and neck brown ; back and wing coverts green with dull golden hues, the feathers marginated with a dusky brown ; greater coverts metallic black and green, waved with a shade of brown and yellow ; tail a ruddy brown ; under parts dusky grey or brown, the legs nearly same colour. The eye is surrounded with a fleshy circle, and has a red streak above it.

History.—This singular fowl was first seen and described by M. Temminck, in 1813 : it is said to be abundant in Java, where it is sometimes seen on the outskirts of the jungles and woods ; but it is as shy as the pheasant.

IV. SONNERAT'S FOWL.

Name.—Coq Sauvage of Sonnerat ; Phasianus Gallus of Latham ; Coq Sonnerat of Temminck ; and Jungle-fowl of the Europeans.

Description.—Size about that of our ordinary dunghill variety, but more slender, graceful, and symmetrical ; from 13 to 15 inches from head to rump ; general plumage a rich and deep ash or slate colour, variegated with maroon and golden green ; the neck hackles and rump hackles a fine maroon marginated with a golden greenish hue ; back and wing coverts of a dark ash, but the shafts of those feathers are of an orange tinge, and have this peculiarity, that they expand in the middle and again at the tip into a flat horny plate, giving the fowl, it is said, a singular and beautiful appearance; the tail, formed of broad flat feathers, with one or two drooping over, of a deep and metallic green; the under parts and thighs a grey or drab; and the legs straw colour, and long; comb large, upright, and indented; bill yellow, with full wattles beneath; throat naked. The hen is about two-thirds the size of the male; of far more sober-tinted plumage; being mostly of a tolerably uniform brown, shaded with a deeper hue, presenting on the back, wing coverts, and wings, the appearance or markings generally known among poultry fanciers here as " partridge; " the under parts a light ash or deep grey; throat and upper portion of the neck white; has only a trace of a comb or wattle, but has a red patch on the cheek, extending beneath the eye; legs of a deep grey. Jardine says that the female has not the horny expansions visible in the shafts of the feathers of the cock, as already described; but Col. Sykes (who wrote from actual observation) states in the proceedings of the Zoological Society, 1832, that their occurrence on the female is by no means rare.

History.—This fowl was first discovered by Sonnerat, and described by him in his Voyage to India; of which continent it is an aboriginal: it is found in the woods and jungles of the western Ghauts and the Mahrattas. It is found also at an elevation of 2000 feet above the ocean. They are excessively shy, wary, and cunning; and therefore exceedingly difficult to take. Their capture is a source of much amusement to the sportsmen of India, and of profit to the lower caste of natives, who snare them and sell them for a livelihood. Though capable of being domesticated, they are uncommonly pugnacious, for which quality they are much sought after by the Mussulmen for the cock-fights, of which they are so fond; and it is said few other game fowls can stand up against them. The Zoological Society of London have frequently exhibited specimens received direct from India.

A fine and genuine specimen of this noble fowl is now (or was some short time since) in the possession of Dr. Horner, of Hull; who states that he obtained it from the Rev. E. S. Dixon, of Cringleford near Norwich,—and that now the Regent Park specimen is dead, his is the only true living cock-bird in this country. " There are," says the Doctor, " some spurious things to be met with, but in those the true marks of the Sonnerat are wanting."

V. BANKIVA FOWL. (*Plate, fig.* 1.)

Name.—Poule Bankiva of Temminck; Javan Cock of Latham; Gallus Bankiva of Lin. Soc.; Ayam Utan, or Brooga, of the Malays; and Java Jungle-fowl of Europeans.

Description.—Size rather larger than our own breeds of Bantam, (undoubtedly descended from it,) and standing rather higher on the legs; neck rather longer, and straighter also; general plumage not much unlike some of the English game varieties, but more brilliant; neck hackle, saddle, and rump hackle, a mixture of rich and deep golden yellow red; beneath the neck hackle blue-black; the back and lesser wing coverts a dark reddish brown or chestnut, and having the feather webs separated; greater coverts metallic or steel blue and black, with an ample arched tail of the same colour, mingled with beautiful greenish hues; under parts shining black; legs pale yellow; head and neck golden orange, but a bare patch at the throat, partially covered, however, by two pendant wattles of moderate size; a scarlet patch on the cheek beneath the eye; the comb large with very deeply indented ridge, presenting almost a "spriggy" appearance; bill of moderate length and dark colour. The hen is said to be rather smaller than the male, and very closely to resemble (in colour only) the brown game hen of this country.

History.—This fowl is a native of Java, in the wild jungles of which it abounds; it is also found in the Malaccas, and indeed may probably be met with in most parts of the continent: Sir W. Jardine speaks of having seen specimens of another species " closely allied, but certainly distinct," also from India, and rather larger than the Bankiva : but his own description leads us to the opinion that they were only *variations* of the same species, perhaps caused by an accidental cross between the Bankiva and some other wild jungle-fowl. Crosses or hybrids are easily obtained between the Bankiva and English Bantam breeds, and many pleasing varieties of this nature have frequently been shown in the gardens of the Zoological Society : and Mr. Martin states in his little sketch, that he " can testify " to the fact of the offspring being fertile,—which, however, we must decline to do.

CHAPTER V.

DESCRIPTION OF THE DIFFERENT SPECIES OF FOWLS, CONTINUED.

HAVING briefly noticed such of the various wild herds of the gallinaceous group as Europeans have been enabled to become acquainted with, we next enter upon the task of describing the individuals which constitute the domestic race of poultry. On no point connected with this subject does there exist greater diversity of opinion, than has been elicited by the inquiry,—to what can the origin of the common fowl be traced? Most commonly some of the wild species of the East are pointed to as the aboriginal stock: whilst a totally opposite theory obtains among many who have given their attention to the matter, and who assert that all attempts to trace the wild origin of our domestic species must fail; that the races of domesticated animals *never were wild;* that the futility of all efforts hitherto made to tame or domesticate existing wild species of fowls, proves that the domestic race must be aboriginal,—from

the beginning designed and created by a benefi-
cent Providence for the uses of man: others, again,
contend that our domestic fowls, are but the
survivors of extirpated and now extinct races that
once ranged the primeval woods and jungles: and
in support of this last theory, it is urged that the
extinction of wild races has taken place, and is even
now going on around us; the Dodo, (already lost,)
the Bustard, Emu, Capercaillie, and Mallard, (gradually disappearing,) are given as illustrations.

It would not, however, be of any practical use
to consider the arguments by which it is sought to
establish these conflicting theories, especially as
we could have but little hope of satisfactorily
settling the point; and until the researches and
investigations of naturalists and scientific men put
us in possession of further means of information,
we may content ourselves in the belief, that in all
probability it is to the operation of each or all of
the causes or principles assigned, either separately
or in combination, that we are indebted for many
of the existing species of domestic fowls.

We now proceed with our description of the

DOMESTIC BREEDS.

Sub-genus Gallus.

I. COCHIN CHINA.

Name.—Loo-Choos, *Chinese;* Cochin China;
Shanghai; and Ostrich Fowl.

Of these, the two first should alone be retained,—
the first as the name by which the Chinese are
said to designate them,—and the second as that by
which they have ever been known to us since their
first introduction to this country. The cognomen
" Shanghai," recently applied to them by some
writers, who seek to substitute it for that of Cochin
China, should be avoided, as the change, being
useless, serves only to create confusion,—a cogent
reason why poultry fanciers and dealers should
scrupulously resist an arbitrary and capricious
attempt to smother a tolerably well-known appellation, for one more un-English, but certainly
not more accurately applied. The last designation should at once be banished, as it originated in a ridiculous supposition that the construction of the wing was double-jointed, and
similar to that of the ostrich; a fallacy we are
surprised to find again repeated in the last edition
of Richardson's little hand-book. If any change in
the name by which these fowls have been from
the first distinguished is necessary, (a point, however, we cannot admit,) surely the most reasonable
and unexceptionable would be to substitute that
of the country whence imported, about which no
doubt is entertained, rather than that of any particular place, whose claim may be a matter of
dispute : and in this view the term " China
fowl " has been suggested, as being less open to
objection ; it is, however, an incorrect form of ex-

pression, as we might with equal elegance and propriety speak of a *Spain* fowl, an *England* fowl, &c. : but it may be worth considering whether "Chinese fowl" would not be a desirable compromise between the rival names of Cochin China and Shanghai or Shanghae. On the whole, we have preferred to retain for the present the one we have selected, as being that almost universally adopted, and by which it is now most widely known.

Origin.—The inquiry " when and whence was this famous breed conveyed hither ? " elicits a very striking illustration of the confusion and obscurity which surround, not only such and similar facts of ancient date, but even those of a recent period. Here we have a foreign variety introduced here, certainly within the last ten years, and yet the poultry historians cannot agree either as to when or whence it was imported! The editors of the *Poultry Book* assert that they were not imported until the year 1845, grounding such statement on the fact that although the Zoological Society held a Poultry Show in May of that year, and offered prizes for " Asiatic varieties," yet no oriental birds but Malays found their way thither ; and also upon the assumption that her Majesty did not receive her presentation specimens until then. Now the first ground affords only a very slender and negative support to the assertion, whilst the other is altogether erroneous : it was shortly after the termination of the Chinese war in 1843, that

Sir Edward Belcher, on his return to England, presented some of these fowls to our sovereign, by whom they were ordered to be placed in the royal aviary,—a confirmation of which is found in a volume published in 1844,* wherein mention is made of the " Cochin China fowls *lately* presented to her Majesty." Those made their " first appearance in public," at the Dublin Royal Agricultural Society's Exhibition, held in the early part of the spring of 1846, whence they became known, and likewise distributed by the liberality of their royal possessor. It has been stated that Captain Heaviside, of Walthamstow, (Essex,) received a present of some, from a friend, with a letter dated Canton, August, 1842; but in the absence of more detailed particulars, we must be excused for believing that her Majesty's were the first introduced here. The next direct importation appears to have been that by Mr. S. Moody, a gentleman residing at Droxford (Hampshire); which were followed in the same year (1847) by the celebrated originals of the stock of Mr. Sturgeon, of Grays, (Essex,) who obtained them by chance from on board a " Chinaman," lying in the West India Docks. So much for the period of their importation hither. In America they did not make their appearance (at least not *genuine* specimens) until 1846. The exact spot which may be regarded as the one of their original location, cannot, perhaps, be accur-

* *Farming for Ladies.*

ately or satisfactorily indicated. The first import-ations were received from the territory of Cochin China, in lat. 9° to 18° north, and long. 105° east: subsequently imported specimens were brought hither in vessels chartered from Shanghai, a Chinese port situated much more northerly than the former country: and in both they appear to be extensively reared. In the *Poultry Book* pre-viously noticed, it is asserted that the birds never even "saw Cochin China"—a statement appar-ently made in utter ignorance or disregard of the many known instances in which specimens have been imported thence direct, as, for instance, the very first which reached our shores, destined for the royal yard.

In America, also, the principal stocks have been obtained from the same quarter: thus in July, 1846, by Dr. Alfred Bayliss, of Taunton (U. S.), through his nephew; and likewise by Capt. Ben-nett Forbes; in May, 1847, by Mr. Taylor, of New York, in the ship Huntress; also by the Rev. M. Brown, a teacher and missionary to China. The birds from Shanghai appear to be-long to a breed precisely identical with the Cochin China, and when it is remembered that a constant and extensive traffic is carried on between them, it seems by no means improbable that the birds have been introduced among the " celestials," by or from the Cochin Chinese: at any rate, whatever claims may be preferred on their behalf, the inhabitants of Shanghai put forth no pretensions to be considered the original domesticators of that race; which indeed, it is said, are known to them only as birds of *Loo-Choos,* that being the name of a group of isles, which, though tributary to China, are situated nearly 500 miles more easterly than that country, and lie southward of Japan. In whatever quarter they were originally found, they must be now pretty widely distributed, as specimens perfectly identical, or seemingly so, have been conveyed to America from the Sandwich Islands, and to Australia from Singapore; they have likewise been met with in the southern portion of the Malay peninsula; and it has also been asserted that they have, for some years, been known and kept at Constantinople: which latter statement, however, we are not inclined to place implicit confidence in.

Description.—This species is particularly distinguished for its extraordinary size—it is an elephant among fowls: and regarding this point as no unimportant characteristic of the true breed, we are disposed, (with Dr. Horner and some few other eminent judges,) to attach more value to that qualification than is at present by many practical men; not, however, that we would make it a *sine quá non,* provided the other points were exhibited in a bird. The body should be of large, almost gigantic (speaking comparatively) proportions, with strong limbs, but compact, solid, and stout;

and by no means running into lanky, gawky, or scraggy fowls. The length of the body from the base of the neck to the insertion of the tail, being about 14½ inches in a full-grown male bird, and the width at the shoulders from 9 to 10 inches, the back being level and straight; girth of the body varying from 23 to 25 inches; the greatest depth from the back to the lower portion of the breast, being about half that measure. The weight is a point on which we should hesitate to speak too determinately, as there are various circumstances contributing to render any fixed standard of size or weight (but especially the latter) useless as a criterion to judge by; the best specimens we are acquainted with range, the cocks from 11 to 13½ lbs. each, and the hens from 7 to 9½ lbs.; a fair average natural market weight per pair being about 21 lbs.; though of course it is no doubt quite possible and practicable to rear fowls which should greatly exceed the extreme weights given above. A rather high standard for chickens (males) from 4 to 9 months, would perhaps be that they should weigh one pound for every month of their age. The general outline of the form is singularly rounded and smooth, possessing none of those sharp angular points which characterize some other species; altogether it may be better understood (if not expressed) by the popular phrase of " square-built." The carriage upright, and not deficient in boldness, although Mr. Sturgeon main-

tains, nay even contends for, the doctrine that
Cochins should, in their true bearing, as he terms it,
" droop forwards with the hinder parts raised ; "
we cannot for an instant acquiesce in such opinion,
however high the authority on which it may rest.
Unfortunate birds so qualified we have seen
crouching in poultry yards, more resembling
Oxford dairy pigs, than true-bred Cochin China
fowls,—indeed it would be difficult to imagine any
more dull, ugly, stupid, spiritless, sickly looking
creatures than they seem to be. The general
plumage is of almost every shade of colour—white,
grey, buff, red, or black,—but of whatever colour
the bird may be, the feathers on the wings and
about the tail *should invariably be of the same* : they
should be fine in texture, shining or glossy, and
soft and silky to the touch : those beneath the rump,
and around the thighs, under and hinder parts, of a
beautiful fine, soft, downy substance ; thick and
abundant in quantity, though not in the least
degree coarse, or even approaching to what is
termed " feathery," but of the finest possible
quality—a point giving a sure indication of *high*
breeding ; this distinctive and peculiar plumage
is called " *fluff* " by some writers, who first manu-
factured the word, and then gave it a place in
their vocabulary as a " technical phrase," without
considering that even a remote meaning cannot be
assigned to it ; as originally invented, the word
was " *pluff*," but this has been discarded, and the

sooner its successor is treated similarly the better ;
" floss," in the sense in which Botanists employ it,
would be a far preferable word to substitute, and
it would at least possess the merit of being capable
of interpretation: the parts named (especially in
the female) cannot be too well furnished with this
downy plumage, which, like that of the ostrich,
we must repeat, should be free, luxuriant, and
elastic, not wiry or firm. Bill of yellow colour,
stout, strong, and curved, not slightly, as some say,
but well curved on the outer upper part; and
withal short, having also at the commencement a
fair supply of tough fleshy membrane surrounding
it. Comb, bright pinky scarlet, of moderate size
only, (comparatively speaking,) springing from
above the nostrils and extending backwards over
the head toward the neck, where it then slopes
off: it must be single, straight, and erect, with even
and well-defined indentations or tooth-like serra-
tions upon the top or ridge; wattles and face
or cheek-patch of same scarlet colour, the former
pendent and moderately long; the ear-lobe also
crimson, but unusually long, hanging much further forward and down than in other fowls, and it
was from this cause that an erroneous impression
at first prevailed that the wattles were double in
the Cochin China. Both comb and wattle, and
indeed all the fleshy excrescence about the face,
should be *fine* in texture and *thin* in substance,
(such qualities being further and sure indications

of high breeding,) and not a coarse, porous, granulated surface. Eye bold, bright, and full; generally fiery red, but sometimes pearl-like,—by some the latter is preferred, (perhaps simply because it is less common,) but it is said that defective, or at least inferior, powers of vision accompanies that description, a statement we can neither verify nor contradict from our own experience. Head small (comparatively), rather narrow, neatly shaped, and rounded outline. The neck only of moderate length, and carried in an erect, dashing, proud manner: the allowance of nine inches given by some writers is certainly in excess; and we venture to predict that when a more correct knowledge of the breed is diffused among poultry fanciers, birds with *such* necks will disappear from our Exhibitions; as we fully agree with Dr. Horner, shortness of neck is an excellent point in a Cochin China, and should be kept in view by breeders as no unimportant and distinctive characteristic of it; birds with this desideration having a more compact appearance, carrying their shoulders higher and their breast more forward; whilst the long-necked ones have an apparent scragginess at the shoulder, with a deficiency of breast. The body cannot well be possessed of too great ampleness of *width* across the back, or of *depth* from the back to the breast and belly. The wings short and rounded in form, clinging closely to the body; the exterior is very convex; they are placed well up on the back;

shoulders nestling forwards, as it were, among the
feathers of the breast, with the ends of the quill
or flight feathers (which are remarkably short)
buried in the rear plumage, thus leaving more of
the sides and thighs exposed than in other fowls;
and this peculiar elevation of the wing into the
back is more particularly observable in chickens
until about 3 or 4 months old, after which time they
would seem to settle or come down somewhat.
The legs should be inclining to short—not more
than 7½ inches long, and as much shorter as breeding can produce; they should be planted well
apart from one another towards the sides of the
body; of yellow colour on the shanks, with a pinky,
fleshy tinge down the scale-less part, like the stripe
on a soldier's trowsers; this mark has been considered immaterial by some, but we think it will
be found to bespeak health in the fowls: the
thighs cannot be too thick or strong, nor too
abundantly and closely covered with the same
fine downy feathers as the rear and under parts;
this down should be so thickly set as to completely
cover and project beyond the hock or shank joint,
projecting over like a pent-house; the shanks
stout and thick, and covered down the outside to
the toes with feathers of the same colour as the
body, and fitting closely and smoothly to the legs,
like the feathers on the legs of a high-bred Pouter
pigeon, coarse quill feathers being very objectionable; the middle toe much longer than the others,

and the membranes or webs connecting the toes
much more apparent than in other species. " Tail
or no tail?" was long the inquiry of amateurs,
for they had heard such marvellous tales of birds
with no tails, that they began to suspect any
specimens that had even a remote indication of
having so comfortable looking an appendage : the
fact is, although the Cochins have remarkably small
tails compared with their size, their deficiency (if
it ought to be called so) is nothing so great as
has been represented; the tail feathers will not exceed six inches in length, of which some are slightly curved or " scimitar " shaped, and fall over regularly in a slightly twisted manner like those of an
ostrich; they are however enveloped almost to the
very ends by the saddle hackle and back feathers,
which rising gradually one above another sweep
backwards over the tail, the size being thereby
even still further diminished—in appearance at
least : the number of feathers should be fourteen,
seven on each side, and the colour to be preferred
is all glossy black, but some of the best birds will
sometimes throw out mottled black and white,
after the first or second moults : the neck hackle
remarkably long, falling very deeply over the
head and shoulders ; the saddle hackle also very
long and close.

The hen is nearly one-third less in size than
the male; the form being very rounded, matronly,
and domesticated in its appearance; the comb

straight, erect, and slightly toothed, but very small in size, gradually rising backwards, until at the highest part it ought not to measure much over half an inch from the root; the wattles small, and receding backwards: the neck cannot be too short, the allowance of eight inches given in the *Poultry Book* being far too great,—such a length of neck ought to hang any fowl. The legs much shorter than in the male, and placed even wider apart, as if to admit of her covering her eggs better; the whole stern and hinder parts round and wide, and entirely covered with a mass of the flossy, downy feathers already described; the tail not quite so long as in the male, and less perceptible, as it is hidden nearly to the tips by the back feathers, which sometimes rise in rather a sharp angle; the more gradual and slight the rise, the more it adds to the beauty of the fowl. In other particulars there is no difference between the male and female requiring any notice.

Of the quiet, peaceable, enduring, and domesticated disposition of this species we cannot speak two highly; they are easily confined in a small fence only a foot or two high, they have no idea of ever leaving home; they are kind to their young, and though not deficient in courage they seldom or never fight: as an instance of their remarkable tameness, we may mention that we have placed a fine cock (a stranger to us) weighing 12 lbs. upon the palm of our hand, and held it out only a few inches from the ground, but it made no attempt whatever to jump off, until absolutely placed upon the ground.

We have now, with great ampleness of detail, described the peculiar points and characteristics of this interesting species, which has (and still continues to do) attracted so much attention. A careful observance of these several points will enable the amateur to select for himself good specimens, —not that we can insure him *pure-bred* fowls, for unfortunately mongrels abound. Before the eggs became distributed from her Majesty's poultry yard, the Cochins had been crossed over and over again with Dorkings. The American *Poultry Book* gives a description of some sent over by Mr. Nolan, (who, we believe, derived his supply in the first instance from the royal yards,) which turned out the verest mongrels that ever were produced: they were clean-legged, and generally resembled the Dorking; they had long bodies, a large, thin, tapering tail, a long neck, full-size wings, and hardly any wattle. Moreover the fact of a bird being imported direct from China will not always insure their genuineness: on this point the celebrated D. Horner makes the following very sensible remarks in a letter to a friend,—"I possess the invoice (or at least the copy) of my imported birds, though are all *bad*, and perhaps sometimes *spurious*, specimens may be, and no doubt are, sent to us direct from China; the only advantage therefore that imported birds possess, is that they show one at least that they are not of the Dorking cross, *of which the country is full.*"

There is no doubt that even at Cochin China the species are repeatedly crossed with the Malays; Camboja, the capital of Malay, being situated betwixt Siam and the former country, and a constant traffic exists between the Cochin Chinese and the Malays. This will account for the Malay-like race of Cochins that made their appearance when the species first came into notice, and which made some of the unfortunate possessors of then contend for the distinctness of the race, til fairly ashamed of their stock. In selecting Cochin Chinas the principal points to *guard agaast*, are long necks, long legs, long wings, and long tails, to which we must add, above all, *clean* legs. On this latter point we are aware some diffeence of opinion did exist, though we believe it is now pretty well dissipated; and we are als aware that so high an authority as Sir W. Jardine has asserted that feathered legs " are only incidental to domestication and cultivation," in support of which he adduces the true *domestic* Bantm of the city whence it takes its name, as having 'plumed legs," whilst the Bankiva (wild bantam has perfectly clean legs. We may also be told that *feathered*-legged fowls sometimes hatch yung ones with *clean* or *unfeathered* legs; but " nture and truth will out," that instance only show that the

chicks have " thrown back " more than their immediate progenitors, and assimilate more closely
to some clean-legged, though it may be far-re-
moved, ancestor. We can only add that we never
saw or heard of a genuine, true-bred Cochin China
that had unfeathered legs; and without a moment's
hesitation we should instantly condemn such as
spurious in every sense of the word.

It now remains for us to notice the different
sub-varieties of this species, which are very numerous if we take shades of colour for the distinction, and of course we can have no other. In
America colour alone is not the distinction carried
out, but the birds are known and distinguished by
the names of the parties who first imported them:
thus we have " Bayliss' importation," " Forbes'
importation," and a host of others, all of which
are supposed to have some peculiar distinctive
characteristic, though of this we should be very
sceptical. In this country a somewhat similar plan
was at first adopted, and the name of the great
fancier who first reared any particular sub-variety
became identified with it. Thus Mr. Punchard's
stock was noted for partridge and grouse-coloured
Cochins, Mr. Sturgeon's for buff-coloured, Mr.
Moody's for the immense size the fowls attained,
and Mr. Herbert's as white. But this practice is
no longer adopted, and Cochins are now distinguished by their colour only.

The different sub-varieties may be included under the following shades :—

White	Brown (light and dark)
Grey	Partridge or Grouse-
Buff (light and dark)	marked
Red	Black.

White. (Plate I. fig. 1.) This sub-variety was
first introduced into America in 1847, by Captain
Palmer, but they were not known here until 1850,
when some were imported by the Dean of Wor-
cester; but fanciers became acquainted with them
through Mrs. Herbert, of Powick, Worcestershire,
who having received a pair from the Dean, reared
some which were sent to the Birmingham Exhibition held in the same year, when they were immediately sought after with great eagerness, for the
rarity of their colour. Many other importations
have since taken place, but Mrs. Herbert's stock
continues to maintain its high celebrity. The
plumage should be pure white without any tinge
or mixture of lemon colour, which is very usually
found in some of the best specimens, and it is a
point very difficult of, attainment to secure them
without; but in our opinion the tinge very materially detracts from the beauty of the birds. The
comb is small, the wattle very long; the colour of
the legs is more of an orange than the darker sub-
varieties; and the tail a *trifle* more ample and
carried a little more elevated. The originals of

our portraits are in the possession of E. H. L.
Preston, Esq., of Southtown ; they are from Mrs.
Herbert's stock, and have carried off several prizes
at the metropolitan and provincial Exhibitions.

Grey.—This differs from the white by being of
a dingy or grey-white, with a good deal of yellow
in the neck and rump hackles, occasionally pencilled or shaded down one side of the hackle feathers with black ; there is also a considerable admixture of black upon the wings, and sometimes
in the tail. The cock attains an enormous size,
but is very compactly built indeed, with no indications whatever of " scragginess." In the fowls
we have seen, the comb has been single, but rather
larger than the Cochin China's usually is, and so
very deeply serrated as almost to give the appearance of a spiked comb; the bill is also rather
longer ; the head small, and a white ear-lobe ;
legs short, flesh-coloured, and feathered. Some
doubts exist among fanciers as to whether there
really is a genuine and distinct grey sub-variety,—
which many will not admit ; and others consider
them identical with the fowl known as the Brahma
Pootra, (described *post,*) for which indeed it is
often passed by dealers, and which it much resembles in the markings.

Buff.—Under this sub-variety would of course
be comprised all the various shades of light and
dark buff or fawn, lemon, " ginger," yellow, &c.:
these however need not all be described, the ama-

teur may safely trust his eye to suit his fancy without the aid of a book. The pale uniform light
buff has now become the most highly-prized and
esteemed shade of all others; and has now for
some time been sought after at poultry sales by
amateurs with an eagerness almost amounting to
furor. The light buff were first introduced into
America by Captain Forbes in 1848; but here
they were not known or reared until a year or two
afterwards. The general colour must, however, be
pure, clear, unbroken light buff, with not a speck
of dark markings upon the neck; all the wing and
quill feathers, and also those about the tail, must be
of the same colour—clear light buff. " The grouse-
coloured breed of Mr. Punchard are certainly
very fine birds," (says Dr. Horner,) " but none but
the best, finest, purest, and clearest coloured buffs
will do for the future, or be even looked at, there
being now so many of other kinds all over the
country." This was the opinion of no mean judge
in such matters expressed some time since, and the
correctness of it has since been demonstrated by
the world of poultry fanciers. Our portraits
(Plate I. fig. 2) were taken, the cock from one
belonging to Mr. Henry Youell, Great Yarmouth,
whilst the hen was the property of W. C. Rey-
nolds, Esq., of Southtown.

Red.—Under this sub-variety we should include
some of those that are miscalled yellow. The
shades vary from an orange red to a deep red ap-

proaching a chestnut colour. The hackle feathers are mostly a deep orange red, the rump hackle being darker than the neck; the wing coverts are usually black, and occasionally we have seen almost the whole of the wing with black feathers; the tail generally black or mottled. The birds in regard to their *plumage* resemble much the Malays, and viewed as Cochins are, to our fancy, excessively ugly.

Brown.—This colour is commonly denominated "cinnamon" by poultry writers, from some fancied resemblance between the colour of that spice and the lighter shades of the fowls,—though we confess we never could see it. The brown varies from a madder to a deep shade or chestnut: the wings are most commonly darker than the general plumage of the body, being sometimes a rusty black, at others a sort of metallic plum colour, neither of which shades adds in any way to the beauty of the bird; and we should unhesitatingly place this sub-variety along with the reds, and say they are decidedly the most ugly of all Cochins. The light brown hens, being of a plain unbroken colour, and much more uniform than the male, is more endurable, and some persons even fancy they have rather a neat appearance; although in those cases, the impossibility of finding a cock that would have any pretensions to match the hen in colour, is a sufficient objection to the introduction of this sub-variety into the yard of any true fancier. There is a broken, mottled, mealy-looking shade of light brown and white, very ugly, but which has been dignified with the name of " silver cinnamon."

Partridge and Grouse-coloured.—These sub-varieties, (if they are not one and the same,) when the markings are clear and distinct, form a very pleasing and interesting variation in the poultry yard. The cocks are termed " black reds," from the fact of their bodies being black, with red or orange neck and rump hackles ; the back a bright deep red, and the wing coverts the same colour ; down the centre of each of the feathers of the neck hackle is a streak of black, which gives a pretty and variegated appearance. Sometimes (but rarely) the cock will be slightly partridge or grouse-marked on the breast and hinder parts. The hens are much handsomer as regards the plumage than the cocks : they are uniformly marked or spangled all over their feathers ; the grouse-markings being generally clearly and well defined, giving the hens, in the eyes of some fanciers, a very beautiful appearance.

Black.—This sub-variety, though by no means the most sought after or valued by amateurs, is certainly the rarest of all ; that is, at least, there is more difficulty in obtaining specimens of *true* unmixed black than of any other colour : indeed so great has been the difficulty experienced in breeding blacks even from blacks, that many persons

who have accidentally obtained a black chicken
from buff parents, have contended that there is no
genuine sub-variety of blacks, but that they are
the result of chance and accident. But the number of black chickens that have been hatched from
parents of the same colour, will require something
more than a few, or even many failures, to disprove
the veritable existence of this sub-variety. We
have no means of ascertaining by whom they were
imported or introduced, or by whom they were
first exhibited; but it is certain that they are only
very partially indeed distributed among the poultry yards of this country, and (as the colour does
not seem to be much admired) it is not likely that
they ever will be,—unless indeed the vane of the
poultry fancy should alter its position. Good specimens should be of a pure, unmixed, glossy black,
having no golden or brassy shades in the neck or
rump hackle, nor on the wing, which latter should
possess a greenish-black tinge upon the coverts.
The comb and wattle, we fancy, in good specimens
are rather larger than other sub-varieties, though
on this point we do not wish to speak arbitrarily.
The legs are a much brighter yellow than in other
Cochins, and the feathers down the leg, which
should be abundant, must be a good black colour.
The birds whose portraits our artist has delineated
(Plate I. fig. 3,) are the property of Mr. J. S.
Brand, the Hon. Secretary of the Great Yarmouth
and Eastern Counties Poultry Association.—

II. BRAHMA POOTRA.

Name.—Brahma Pootra; Grey Shanghai, or Cochin; Chittagong (Americanism).   The first would seem to be the correct name; the second being only applied to it by those who imagine this breed to be a sub-variety of the Cochin.   With regard to the last name, the New England Society for the improvement of Domestic Poultry made the following observations upon some entered under that designation, at the Show of 1851,—" Mr. Hatch's lot was entered under the head of *Grey Chittagongs*, but were really pure Brahma Pootras. As the judges desire that every variety of fowl should be called by its right name, they cannot sanction the application of the title Chittagong to this excellent stock, when in reality they are perfect Brahma Pootras."

Origin.—This breed would seem to have been first imported into America (whence they originally became known) in the latter part of the year 1850. As the tale is told in the *Northern Farmer*, (a paper published in Oneida County, U. S.,) some sailors imported a few from the banks of the Brahma Pootra, the largest river in India, and known in Thibet as the Sanpoo River.   The Editor goes on to state that American fanciers are as yet divided as to whether they are entitled to be considered a distinct breed, or only what they term a superior sub-variety of Grey Cochins: others, with Dr. Bennett (author of the *American Poultry Book*) at their head, affirming their conviction that even apart from the consideration of the widely separated localities in which the respective breeds have their origin, they possess sufficiently well marked characteristics and points of diversity to entitle each to be considered a distinct breed. Mr. Nolan, of Dublin, writing recently to the *Cottage Gardener*, states that he has received some Brahma Pootras from America, and that he thinks them " distinct," but that they " are a sub-division of Cochins,"—a piece of contradiction that needs not be pointed out. The first importation of the breed to this country, appears to have been some sent to her Majesty as a present from Dr. Bennett: at the same time some were sent to Dr. W. C. Gwynne, of Sandbach, Cheshire, and Mrs. Hosier Williams, of Eaton Muscott, near Shrewsbury. The shipment was from Boston, and took place in the year 1852. At the Metropolitan Show held in July, 1853, there were three pens only of these birds exhibited, and on these the Editor of the *London News* remarks,—" There is a class of fowls that seem likely to outrival even the Cochin Chinas themselves: they are the Brahma Pootra fowls. Not only with regard to superior quality of flesh, but from the quantity of meat they have on the breast, they are considered to be superior to the Cochin China." Of the three pens shown, that of Dr. Gwynne carried off the prize. At the Great Yarmouth Exhibition, August, 1853, there were but five pens, (a proof of their scarcity,) the prize being awarded to J. Fairlie, Esq., of Cheveley Park, Newmarket.

Description.—The size to which these birds attain is very great; indeed, some writers have gone so far as to assert that they get as large as Turkeys; a statement we should be sorry, however, to venture on. Their form is remarkably round and compact, more so than the Cochin Chinas, and they are much deeper and fuller breasted than that variety, but shorter quartered; they have also the advantage of being shorter legged and shorter necked. The average natural weight per pair is said to be from 22 to 25 lbs.; the cock ranging (according to some writers) from 11 to 15 lbs., but the last we should say is a very extreme weight. The bill is yellow; rather more curved on the upper ridge, and stronger, than that of the Cochin China; the comb usually what is termed a " pea " or " bean " comb, and placed almost on the forehead, the long feathers of the neck hackle commencing as it were on the crown of the head, and immediately at the termination of the comb; occasionally the comb will come out single, upright, and serrated, although in this case it is very small; but the pea-comb is the most admired in America, and undoubtedly it is the most to be desired, as presenting a greater distinctive mark between them and the Cochin Chinas than the upright combs. The wattles are small, but the ear lobe is large and pendulous, both of a bright crimson colour; the small head and short neck contrasted with a broad, ample chest. Wings very short and rounded; under parts, like the Cochin China, profusely furnished with fine, glossy, downy feathers; legs short and flesh-coloured, and heavily or thickly feathered down to the toes; tail small and of fine glossy black colour, with sometimes a few dark green plume feathers peeping above the others. The general plumage is white; the neck hackle of a lemon hue, each feather being very beautifully and regularly streaked or shaded down with black; the wings are also generally slightly shaded with black; the rump hackle of a deeper lemon or yellow shade than the neck hackle, and frequently free from black streakings; though of course more admired with shadings. The hen is at least one-third less than the male, with a small head, scarcely perceptible comb and wattle; neck short, but upright in its carriage, with the feathers of the neck of a lemon shade, and streaked with black, though the markings or shadings are seldom so beautiful as in the cock; the back and wings also similarly shaded, only slightly; tail black, and carried in a more elevated position than by the Cochin Chinas.

Our portraits (Plate II. fig. 1) are from fowls

the property of Mr. J. S. Brand, of Great Yarmouth, bred from Dr. Gwynne's stock.

III. MALAY.   (*Plate* III. *fig.* 1.)

Name. — Malay; Chittagong; Domesticated Kulm Fowl, or Gigantic Cock, of some writers; and Java Fowl.

Origin.—This is another breed of Asiatic origin; at first introduced from the peninsula after which it is named, and which is situated on the extreme southern point of the Indian continent. They have long been domesticated in this country, and probably in Ireland also; although Mr. Nolan of Dublin claims to have imported the first true Malays, which he purchased in the London Docks, they having been brought direct from the Malay peninsula. Some years since frequent importations of these fowls took place at Falmouth, out of the East Indiamen that touched at that port. It is said that the breed is also met with in the Isles of Sumatra and Java; but it is tolerably certain that they are not natives of those places, but have in all probability been introduced thither by the Mussulmen. In the *American Poultry Book* mention is made of the following sub-varieties,— "Bucks County," "Jersey Blues," and "Boobies;" but as no account of their origin is attempted to be given, we strongly suspect that they are nothing but crosses, of which our friends on the other side of the Atlantic are remarkably fond, as a means of exercising their experimental ingenuity. The same work also contends for the Chittagong being a *distinct* breed from the Malay, contrary to the opinion almost universally prevalent among poultry fanciers: and mention is made of a breed of Chittagongs, with grey body, the back, wings, and hackle feathers of a "silvery-yellow," with stray feathers of brown and white ; a single comb, with sometimes a "top-knot," and feathered legs. It seems hardly necessary to remark that such a description of a fowl bespeaks at once its mongrel origin : and notwithstanding the formidable array of sub-varieties named by brother Jonathan, we are still disposed to assert that there is but one true variety of the Malay fowl, and that it is identical with the "Chittagong" breed of poultry writers.

Description.—These birds are of gigantic proportions, and attain to a remarkable size ; they stand 27 to 28 inches high, and can readily pick food at a height of nearly three feet from the ground, and before the introduction of the Cochin Chinas they were undoubtedly the largest inhabitants of the poultry yards. The form is remarkably long, and the body slopes down at a very sharp angle from the bottom of the neck to the commencement of the tail : the weight of the cock is from 9 to 11 lbs. (generally), although the prize bird at the Birmingham show in 1852, weighed 11¾ lbs. ; and, in a letter to the Royal Dublin Agricultural Society, Mr. Rutherford states that he had one, *unfatted*, that weighed 13 lbs. ! These, however, are extraordinary weights, very rarely attained. The hen averages in weight from 7 to 9 lbs., but the prize one at the above-named exhibition weighed $10\frac{1}{4}$ lbs.: the bill is short, strong, and curved, of a yellow colour; the comb of a very peculiar form, singularly stunted and depressed, extending scarcely so long as the top of the head itself, but coming well forward on the forehead,—it is sometimes described as a double comb, cut short off and flattened, but this conveys only a very poor idea of the correct form and appearance; it looks in shape on the top as if it consisted of four or five small horse-beans enclosed beneath a bright red skin very tightly stretched over it, so as distinctly to show the form of the beans; indeed the comb is not unfrequently now described as a " pea or bean comb. " Occasionally we have heard of Malays with small single upright combs, but we should be inclined to regard any birds as not true-bred that did not possess the form of comb we have described above. The wattles are almost entirely wanting, but there is a quantity of loose red skin which surrounds the cheek as if to supply its place: the eyes are fine, large, and bold, and piercing in their glance, usually red with the iris of a bright yellow; but sometimes the eyes are what is termed " pearled " round the edges with almost a white colour: the head is singularly neat and clean-shaped, formed, as it has been constantly de-

scribed, like that of a " serpent," sleek, depressed
on the top, and elongated : the neck is extraordinarily long, sometimes reaching 10 inches, and
owing to the feathers of the neck being very close
set, and not being more abundant at the bottom
(as is commonly the case in other fowls) than at
the top, the neck appears almost entirely of a uniform thickness of about $6\frac{1}{2}$ inches round ; the throat
for about an inch beneath the lower bill is entirely
destitute of feathers. The general plumage is of
a very hard and close texture, the colour being a
rich brown or dark chestnut, darker much on the
breast, back, and thighs : the wings long and projecting downwards to the ground, the coverts of a
rich glossy metallic black, the quill feathers lighter,
that is, about the same shade as the feathers on the
breast ; the under and hinder parts much resembling those of a trimmed game cock : the legs are
of a light yellow or straw colour, with a pinky or
ruddy streak down the side, something like the
mark on the leg of the Cochin China, and which,
as we before observed, we take to be a sign of
health ; both the shanks and thighs are of most
monstrous length, sometimes falling but little short
of 12 inches ! the bones of the legs are very stout
and strong ; and the skin is covered with large
though not coarse scales : the tail flowing and
erect, (but neither large nor long,) with some few
beautiful sickle feathers falling gracefully over,
and tapering gently to points, generally of a black

colour, with dark blue and green metallic reflections in it: the neck and rump hackles, of the variety above described, a beautiful deep rich brown, the feathers being extremely glossy and shining.

The hen is of course much smaller than the male: the same bill and comb, though the latter is smaller; no wattles, but a bright red skinny face and cheek; the same " serpent " head; the neck very little shorter than that of the cock, and carried in a similar erect and dignified manner; the wings long and projecting downwards to the ground in a much greater degree than in any other fowl; the legs of the same colour and very nearly as long as those of the male; the tail feathers (said to be five in number on each side) very straight and carried erect, overlapping one another like a fan, and projecting one beyond the other so that the last, or upper feather, is the longest: the colours of a hen to match the cock described above would be, the wings and back of a deep brown, the neck and rump hackles of a lighter or reddish brown, and the feathers of the breast and under and hinder parts of a still lighter shade; the tail feathers black or very dark brown.

Sub-varieties.—These are more numerous than might be supposed. The breed commonly known some years since appears to have been a rich deep *red*, striated or streaked with orange or yellow; the back an orange red; neck hackles deep ruddy brown, somewhat of a lighter shade than the feathers of the rump hackle : this sub-variety we think the neatest and handsomest of all. There is also, though rather uncommon, some *light browns*, (the dark we have already noticed as being the most widely spread,) with occasionally a few yellow streaks in the male, and a light chestnut-coloured breast, deeply marked or spangled with black ; but the hen Mr. Baily (the eminent poulterer) describes as being one whole uniform colour,—a light brown or chocolate shade, with occasionally the feathers of the hackles somewhat darker than the rest of the body : these we have never seen, but can readily imagine that they must have a very pretty appearance indeed. The same gentleman also speaks of a remarkably handsome sub-variety coloured " like game piles." There is a very beautiful sub-variety of *white* Malays, the plumage of which is " snowy white " except the feathers of the hackles, which are strongly tinged with a lemon or light yellow colour ; it is asserted by some writers that the whites are smaller in size than the common brown Malays, but we know not whether this statement is made from actual observation of the fowls, or only hazarded on a speculation to appear well acquainted with them,—probably the latter is the case, as we have met with several similar instances in our readings in poultry literature. There are also some hens we have seen nearly black, streaked with brown

and yellow rather prettily ; but have never seen
a cock that would match them, and they have
been placed with red or brown cocks. The Rev.
E. S. Dixon, in his work on Poultry, mentions a
perfectly black sub-variety of Malays, which he
says are very "handsome and cavalier-looking
birds ;" and a correspondent states that he saw a
lot of them in Hungerford market, which he was
at first inclined to think had a cross with the
Spanish, but having recollected that he had seen
fowls of exactly the same appearance in Devon,
he changed his opinion. We are unacquainted
with this sub-variety, but should think with the
above gentleman, that it would form a handsome
addition to our domestic poultry. The Malay
breed is said by some to possess game blood, and
most assuredly their generally pugnacious inclination and readiness to encounter any antagonist in
the yard, would seem to confirm the opinion. The
breed many years ago was much more prized than
it has been for some time, and we greatly fear that
it is fast disappearing in this country,—a matter
to be regretted, we think, by all true poultry fanciers. At the Poultry Show held at Birmingham,
1852, there were but ten pens of Malays sent,
at the Metropolitan Show in 1853, there were
eighteen pens, and at Great Yarmouth, 1853, only
five pens. We trust, however, that the encouragement offered to breeders by these Exhibitions,
may have the effect of bringing again into notice

and favour so distinct and handsome a variety as the Malay really constitute.

IV. GAME.   (*Plate* III. *fig.* 2.)

Name.—Game Fowl; Coq D'Angleterre of Buffon; Spanish Game Fowl (American).

Origin.—The progenitors of this once famous breed, styled *par excellence* the "English Fowl," it has been generally assumed, were introduced into this country by Julius Cæsar: but the account he gives of his first visit, (to which reference was made in the earlier pages of this volume,) shows that even at that early date, and perhaps long before, the inhabitants, though avoiding their flesh, kept fowls for their pleasure and amusement; and it is difficult to imagine what other amusement could be afforded to the barbarians of those times, than that for which, down to a recent period, they were so much sought after by their more civilized descendants,—namely, the exciting and sanguinary conflicts in which they were trained to take part. Some have sought for the original of the breed in Eastern climes, among some of the wild fowl of the jungles: and although they may, doubtless, have in modern times been crossed with foreign varieties, that would not seem sufficient ground for rejecting the presumptive evidence we have for supposing the breed to be aboriginal inhabitants of our island.

Description.—The Game Cock, viewed as an ornamental fowl, far excels, in our fancy, any other of its species, whether we regard its light and elegant form—its graceful and majestic carriage—its bold, proud, and courageous bearing—or the brilliancy and beauty of its plumage. It has not inaptly been likened in its relation to the other inhabitants of the poultry-yard, to the race-horse among horses, or the greyhound among dogs. The size is somewhat below the common, though inferior specimens may often be met with which are large and even unwieldly in size; the form is elegant in outline and perfectly symmetrical in every part; weight of the male about 5 lbs. to 6 lbs., and of the female from 4 lbs. to 5 lbs.: head small, narrow, and elongated; comb (before " dubbing " or cropping) of a moderate or medium size, fine, and usually of a bright red; wattles and cheeks of the same colour; the eye large, full, and bright, presenting the appearance of a sparkling glass bead; the bill curved and very stout; neck long and sleek, but full; breast very ample and broad; wings large and strong, of a convex or well-rounded form on the outer sides, and drooping downwards to the ground, covering the thighs, and, when well developed, projecting rather below the body, which is very round and compact, but tapers gradually towards the tail; thighs strong and firm in muscle, the shanks powerfully set on and inclining to be long; feet flat with full-size toes, armed with strong claws of by no means despicable dimensions : the neck and rump hackles finely feathered, long, and abundant; the former reaching to the back, which should be broad at the shoulders, flat, and sloping : the tail thickly set with feathers at its insertion, and falling over gracefully in what are called "sickle" or semi-circular shaped feathers. To please "the fancy," the game cock must possess many points besides those we have described : the plumage must appear to the touch very hard, close, and compact in its texture, very closely resembling that of the Malay, (to which breed it has been fancied the game fowl may be distantly allied,) and the whole body must be in exact proportion ; and to such nicety is this latter point insisted on, that if a fowl should not exactly balance itself, when placed on the hand of a fancier, supported just beneath the breast and before the legs, it would undoubtedly be rejected : another and still more important requisite, however, is that the spurs should not be seated too high up on the legs, in order to enable it to give more effective blows in a conflict with a rival.

The hen preserves the proportionate difference in size between the male, usual with this tribe of birds, namely, about one-third smaller, though equally neat in figure and appearance : the comb is small and upright ; the limbs are flexible, and yet muscular ; in other points she does not materially differ from the cock,—if we except the tail, which is large in all the sub-varieties, fan-like in form, and much elevated above the back,—the tip of the uppermost feather being very little below the level of the head.

Sub-varieties.—In our remarks upon the nomenclature of fowls, we have spoken of the evils resulting from the "fancy-naming" of poultry; and had we needed an illustration in confirmation, the game breed would have supplied it,—for assuredly never were any other poor birds so cruelly knocked about with original and home-manufactured nick-names as they have been! We have collected, out of various works, a list of names of no less than *fifty-three* (supposed) different sub-varieties or "breeds," the whole presenting a most unintelligible and bewildering array of names, (in many instances purely *local*, and in others even *slang*,) and the multiplicity of which is alone sufficient to cause the greatest possible confusion. Much of the mystery which surrounds the history of its sub-varieties, has been occasioned by the injudicious application of separate and distinctive appellations to the numerous progeny which have resulted from the system of crossing the various "strains," (no matter how opposite may have been the colours of the plumage,) which has so long been indiscriminately pursued with regard to the game fowl.

Without assuming any claim to infallibility, we trust that the subjoined list of sub-varieties may be found to be both complete and convenient in arrangement:—

WHITE, or Smock (vulgar).

GREY.

Clear mealy or dun colour.
Black-breasted, or red dun.
Dark grey, blue, or smoky dun.
Cuckoo.

RED.

Dark, or " Ginger."
Brown-breasted.
Black-breasted.
Muffed (or bearded).
Tasselled (or tufted).

BLACK.

Pure black.
Brassy or copper-winged, and " Worcestershire Red."

Golden-hackled.
Furness or Polecats.

PILE.

Cheshire.
Staffordshire.
Worcestershire.
Henny.
Red streaky-breasted.

DUCK-WINGED.

Silver-breasted, or birchen grey.
Yellow-breasted, or birchen yellow.
Black-breasted, or dark birchen.

HENNIES, or Hen-cocks.

White.—Fowls of this colour are difficult to meet with, and appear to be almost exclusively confined or located in the midland counties; where, however, they are well known and highly esteemed as one of the most beautiful of its species. It is said to be an importation from the East; but it is the opinion of those who have had much experience in breeding them, that they are as purely English as any of the other sub-varieties. The white should be uniform and clear,—free from all yellow tinges. The legs are of a rich creamy colour, or yellow, but the former is preferred, as preserving better the uniformity of colour.

Grey.—Is of what is called a " clear mealy " or dun colour, streaked with dark or black feathers upon the breast and hackles, the ground colour of the latter being a light yellow or straw.—The *Red dun* differs only from the last in the mealy or dun colour being broken with a reddish-brown upon the back, wings, and hackles; the feathers on the breast being also much darker, approaching a rusty black. The *Blue* or *Smoky dun* is by far the handsomest of its kind; and is better known in the county of Dorset than in other parts of this country, though it is but recently that they have been admitted into the Game family, having long been looked upon as a distinct breed—which they most assuredly are not. Their colour is a rich, soft, slaty blue; the neck hackle and wings darker than the rest of the body, but sometimes variegated with streaks of yellow hue or dark red; the tail has often the long drooping feathers barred or striped with black.—The *Cuckoo* Game fowl much resembles the grey-barred Dorking, (sometimes called " Cuckoo Dorking,") its barred plumage presenting the same shades as are to be found on the breast of the cuckoo: it is, however, neither admired nor esteemed, and is now fast disappearing.

Red.—The dark red (or " Ginger," as it is sometimes most inaptly styled) sub-variety, is that most commonly met with in the farm-yard. The comb is rather larger than in other breeds;

the breast of a reddish or rusty black; the neck hackle of a deep orange red, as also the rump hackle; the back and wings of a deep brown or chestnut colour; the tail, a rich metallic green and black; and the legs white or very light colour: the hen is of a sober brown plumage relieved with a few yellow or ochre-tinted streaks.—The *Brown-breasted* is smaller and neater formed than the last: the neck and hackle is of a golden-red or orange colour; the breast a fine and deep brown, richly spotted or spangled with golden hues; the wing coverts also spotted; the back a rich red shade; the tail fine glossy greenish black; and the legs a dark leaden colour.—*Black-breasted* Reds are by some amateurs held to be the veritable originals of all the other kinds of game fowl: and they are the breed most esteemed, by those who pretend to the " fancy," as the best specimens that remain of the true old English game cock. Of this sub-variety, the purest and truest bred are those known as the " Derby strain," having been most extensively reared by the late Earl, at Knowsley,—where they have been kept for upwards of a century; the greatest care having been taken, during the whole of that time, to prevent the slightest deterioration in the stock, for which purpose they have at occasional intervals been crossed with the strains in the possession of Lord Sefton, Mr. Germain, and other fanciers. The general plumage of the cock bird is exceedingly showy: neck and

I

rump hackle of a bright, deep golden-red; the
back a clear, unbroken, dark red; wings, a rich
reddish-brown or a maroon colour, the greater
coverts being marked with a band of purple or
steel-blue shade across it; breast perfectly black,
so also the thighs, which are very stout; the shanks
of the leg a white colour, and claws the same; the
comb larger than in many sub-varieties, and a fine
red cheek-patch; the eye a clear, fiery, pearl-like
bead, resembling the eye of the jackdaw, and
hence this very peculiar organ is sometimes designated a " daw eye;" another characteristic of
the true-bred fowl is a striped or streaky white
bill; the fine tail is of a beautiful lustrous black
colour, with metallic greenish shadings.   The hen
is by no means so brilliant in feathering as the
male: the neck hackle is a pale reddish brown,
shaded and streaked with black and yellow ochre;
breast black-edged, with buff or ochre shades;
under-parts of a dull slaty grey colour; the back,
coverts, and wings much resembling the colours
and markings of the feathers of the partridge;—
the quill feathers of the wings, however, are black,
and the tail of the same colour.—The *Muffed* or
bearded, and the *Tasselled* or *tufted*, game, scarcely
deserve separate notices, as they are so seldom
found to occur, that they would seem rather to be
the result purely of accident; and moreover they
are not *exclusively* met with among the Reds: the
first above-named owe their designation to a small

ruff or beard beneath the throat, differing, as far as we can learn, in no other particular from any other fowls of similar colour: and the second, or tasselled sub-variety, are so called from their having a few irregular stray feathers sticking out most inelegantly behind the head in the form of a small tuft or tassel.

Black.—Game fowls of a *pure unmixed* black plumage are, like the true black Cochin Chinas, exceedingly difficult to find: they being mostly streaked, more or less, with golden yellow upon the hackles and wings, the different markings, or gradations of shade, being designated by some peculiar name.—Thus we have the *Brassy-winged* Blacks, differing from the black only in having their wings barred with a yellow, or "brassy," band; the shoulders also sometimes being marked with patches of the same: the hen is generally a good black, with a brown breast: these birds are also known (and more aptly described) as "Copper-winged" Blacks; and at the Birmingham Exhibition of 1852, a pen belonging to Mr. J. T. Wilson, of Redditch, Worcestershire, was entered as "pure Worcestershire Reds."—The *Golden-hackled* are the Blacks having only the neck hackles streaked with yellow or brassy shades, and, unlike the preceding, having the rest of the plumage unmixed.—The *Furness*, so called, it is said, from a parish and hundred in Lancashire, in which county it is highly prized: its plumage is a bright lustrous black, except only on the saddle, which is of light buff, sometimes approaching a golden or copper colour : the hen is a very much better black than the females of most of the other sub-varieties ; her neck only being streaked with brassy shadings. The *Polecats* closely resemble (and most probably are the same as) the Furness, from which it can be distinguished but by some few streakings upon the wings, of the same colour as the saddle. The Furness fowls are frequently vulgarly called *Furnace*, from a fancied resemblance of the shades on the saddle to the glare from a heated furnace. The Brassy or Copper-winged Blacks are sometimes incorrectly called Furness fowls.

Pile.—This term is universally applied by fanciers to those game fowls, of whatever colour the general plumage may be, that are *pied* or variegated with white feathers ; and it seems highly probable that the name is nothing but a corruption from *pied.*—The *Cheshire Pile* is a white fowl with an unmixed deep red back, and the breast streaked with the same colour.—The *Staffordshire Pile* presents the same variations of colours and shadings, but the red is of a light orange or carrotty colour ; and the wings are not unfrequently streaked as well as the breast.—The *Worcestershire Pile* is indiscriminately applied to fowls of the same appearance as the preceding, but having the white, or yellow, or cream shade sparingly variegated with black markings in addition to the red,—and also to the *blues* or *greys ;* in which the only difference consists in a slaty-grey shade being substituted for the yellow or cream colour. The *Henny pile* was many years ago well known in Cornwall and the adjoining county of Devon ; and owed its name to a remarkable similarity of form and appearance existing between the cock and hen : they are supposed by some to be a sort of sub-variety of the Hennies which will be afterwards noticed.—The *Streaky-breasted red pile,* which is very often inaccurately classed among the Red Game fowls, most closely follow the plumage of the Brown-breasted Reds, previously described ; the only difference being in the colours of the wings and tail, which in the Streaky-breasted are prettily mottled with white, after the manner of the Pile or pied fowls.

Duck-winged.—This sub-variety (so named from the metallic lustrous shades on the wing, like those of the duck) is generally admitted to be the most handsome of the game kind, and equalled only by the celebrated Derby for other points and qualities : the colours are exceedingly varied, and though brilliant often, they are beautifully and harmoniously blended.—The *Silver-breasted,* (sometimes called mottled-breasted,) or *Birchen Grey,* have the breast and under-parts white ; the neck hackle and saddle grey, streaked with brown or black ; the back and shoulders a deep orange or red; the wings marked with that peculiar metallic shading which characterizes this sub-variety, and whence the name is derived; the tail a rich black with greenish reflections.—The *Yellow-breasted*, or *Birchen Yellow*, have the breast of a deep yellow ochre streaked or mottled very beautifully with black; and the same shade of ochre prevailing more or less throughout the rest of the plumage.— The *Black-breasted*, or *Dark Birchen*, is the most esteemed of the Duck-winged fowls: the breast and under-parts are of a deep black, and should invariably be unbroken with any other colour; the neck hackle is of a light straw-colour ground, with grey and black markings and streaks; the back a deep red or even bright chestnut colour, gradually fading through orange and coppery shades to a bright golden tint in the rump hackle.

The hens of all the kinds of the Duck-winged sub-variety, though neat and pretty in their appearance, present but few of the points of difference in colours, which have divided the males into classes. The legs and bills of both male and female are usually light grey or ash colour in the Birchen Greys—bright straw colour in the Birchen Yellows —and dark slate colour in the Dark Birchens.

The *Hennies*, or *Hen-cocks*, as they are called by some, are a very singular but nearly extinct sub-variety; the great peculiarity of which consists in the male and female bird being as nearly as possible identically alike, both in regard to the shape and plumage : the latter is of a dusky brown, with black and buff streaks and marks, not much unlike what is termed "partridge markings," though scarcely defined so well.   The cock has a very small and straight-feathered tail, and being without any of the long hackle feathers on the neck and saddle which usually distinguish the male birds, there is no perceptible difference between him and his mate, in appearance at least, and hence their name.   These fowls are but little known, and in the absence of further information it is impossible to say whether they are a distinct breed, or only the result of crossing,—the latter being by no means unlikely.

V. DORKING.   (*Plate* III. *fig.* 3.)

Name.—Dorking Fowl; Gallus pentadactylus of Temminck; Coq a cinq doigts of Buffon ; and Sussex Fowl.

Origin.—This fowl, so called from a small town in Surrey, where probably the variety was first systematically and extensively reared, (being found there in greater purity and perfection,) is undoubtedly a breed of great antiquity, having been noticed and described in the first century of the Christian era both by Columella and Pliny ; and there seems fair grounds for supposing that these birds were introduced into this country by the

Romans, among whom they had attained, at that early period, some celebrity, and were much esteemed: with us but few fowls can boast such high and long-continued reputation as the Dorkings. It has been suggested that Shakspere was acquainted with the superior qualities of these fowls, and that he alludes to them in his Henry IV., when he makes Justice Shallow, " of Glo'ster," order " a couple of *short-legged* hens " for his guest's repast.   The chief distinctive mark or characteristic of the breed is the presence of a fifth or supernumerary toe, (more properly speaking claw, as the joints are very rarely indeed articulated,) springing behind, a little above the foot, and below the spur.   It has been sought by various writers to deprive Dorking of the honour of being the original and principal rearing-place of this justly celebrated variety : and it is asserted that the true Dorking Fowls are raised at Horsham, Cuckfield, and other places in the Weald of Sussex, and bordering upon the county of Surrey; and that the ancient and superior white fowls from Dorking are a degenerated race compared with the " improved " Sussex breed; that their having five claws is by no means the true and original characteristic,—such peculiarity being merely fortuitous, and in fact even objectionable; and that those so marked are deemed a mongrel or bastard breed. With what evidence or argument it is attempted to support these assertions, it would be a waste of time to inquire: it is sufficient for our purpose that we possess such a variety, and know where to obtain it in the greatest perfection. No doubt it is probable that their five claws might have, in the first instance, accidentally brought into notice certain fine and well-formed individual birds, but from these proceeded a distinguished and permanent variety; and that variety bearing the name of Dorking seems to be a sufficient, or at least a presumptive, proof in favour of that town and its neighbourhood. In the mean while, the appellation of Dorking Fowl has been in use, we apprehend, far beyond the memory of any individual living; for upwards of a century ago a writer in the *Gentleman's Magazine*, remarking upon the incredible number he saw at Dorking, observes that the fowls (especially capons) were " well known to the lovers of good eating," as being remarkably fine birds; and it is now some sixty years ago since the original author of this treatise, whilst resident in the county of Surrey, sent to Dorking for his first regular breeding stock, and obtained some genuine specimens of the " ancient five-clawed " fowl, at that time in high repute. It is not at all improbable that what is termed the " improved " breed of Sussex Fowl, has originated from a Dorking cross,—the peculiar characteristic mark of five claws disappearing sometimes in the course of breeding, owing to the smaller number of Dorking cocks employed, as compared with the common cocks of the large Sussex breed, which were not so distinguished. Such, it is well known, is very commonly the case in crossing different varieties of live stock : the home or original variety in the end gets uppermost as being the majority. Fowls evidently partaking of the cross above suggested, abound in Sussex, Kent, Berkshire, and many other parts. The true Dorking breed is now tolerably well scattered over England. It was at first rather extensively reared down in Cumberland, whence the breeders introduced it into Lancashire and Westmoreland; but it has been found by experience that, from some cause or other, they do not continue to maintain that high character and quality which they have so long possessed in the immediate vicinity from whence they originally became known and dispersed. They have also recently made their appearance in Ireland.

Description.—Before giving a description of this variety it may be well to premise, that although the subjoined characteristic points may be found, more or less, in all the crosses or sub-varieties generally included under the appellation " Dorking Fowl,"—they are intended to apply exclusively to the *white* Dorking, which, we must affirm it to be our conviction, *is the only genuine and original breed*, entitled to that name : our

reasons for which will be more fully stated in considering the " Sussex Fowl," or alleged "improved " breed of Dorkings.

This variety is inferior to very few others in
size; the form in a great degree approximates to
that of the Cochin China, the body being large,
plump, and square-built, but withal well-shaped,
short, and compact, with broad back and shoulders,
and capacious full breast,—in fact, no other bird
has so little offal, or so much flesh stowed away
in similar compass, as the Dorking has; the
weight of the cock is good if from 7 lbs. to 8½ lbs.,
and that of the hen perhaps 5 lbs. to 7½ lbs.: the
head not large, and rounded upon the top, on
which should be a thick-set *double or rose comb*,
of a fine scarlet colour, and with regular though
small points; the wattles of the same hue, *of moderate size only ;* bill a cream-white shade, curved
and rather blunt at the end; neck very thick,
lumpy, and short, and set upon the body straight
and ungracefully ; wings round, thick, and strongly quilled: the legs very short, smooth, and white
or light flesh-like tint; the foot *furnished invariably
with five toes*, that is, four complete toes, (the hind
one being rather larger than usual,) with an imperfect fifth or supernumerary toe, rising like a
second spur from the same root as the hind toe
springs from in common varieties; the tail well
arched, with flowing and sickle-like feathers:
plumage of the *true breed* perfectly white—un-

broken and entire; the neck hackle long and thick, the rump hackle of moderate length only, but abundantly furnished to the fowl. The hen differs but very slightly from the male, than which she is only about one-fourth smaller: her comb small and thick; breast remarkably round and pouting; the tail straight, vertical, or fan-like; the same number of toes or claws.

This, the genuine Dorking breed, owing to the innumerable crosses to which they have been subjected, is now becoming exceedingly scarce, and can scarcely be met with beyond a very circumscribed district in Surrey. At the Birmingham Show, in 1852, out of 132 pens of so-called "Dorking Fowls," only 19 were true white; and at the Metropolitan Show, in 1853, there were only 22 out of 98 pens.

Sub-varieties.—From our previous observations it will have been collected that we do not admit, strictly speaking, of any sub-variety of the true Dorking Fowl: but as a mixed (not to say bastard) breed has, by numerous crosses during a long series of years, become established almost as a permanent variety, under the same name, it may perhaps be advisable to treat it practically as a *distinct branch* of the Dorking Fowl, which will comprise all the coloured sub-varieties. We shall thus have,—

The Sussex Fowl, *or* " *Improved* " *Dorking.*

GREYS.	BLACK-BREASTED.
Speckled.	Silver.
Spangled.	Golden.
	Japan.
REDS.	
Speckled, or pied.	CUCKOO-BREASTED.
Pencilled.	

Sussex Fowl.—This kind of Dorking is alleged by many to be an improved breed, but by some fanciers they are held to be the reverse; in America the old white stock maintains an undiminished reputation. Dr. Bennett, the author of the *American Poultry Book,* speaking of the Sussex Fowl, says, " They are considered by some to be an improvement on the original Dorking; but in my opinion they are not equal to the pure white, imported direct from Dorking, in Surrey, England." On the other hand, Mr. Baily (certainly no mean authority) gives his decided preference to the coloured Dorkings, or Sussex Fowls, as being " better for the table," (a matter we shall subsequently discuss,) and hardier than the white; but he admits that the former sometimes throw *four*-clawed instead of *five*-clawed chicks, and although he does not consider it of much importance in home-bred birds, still he says he would " always insist on the presence of five claws " for stock birds,—a sufficient reason why those who deem it a characteristic of *true breeding* should not look for it among the "improved Sussex breed!" Mr. Tegetmeier, in his little pamphlet, after extolling the dark-coloured family, complains that there is "*a difficulty in breeding true to any markings:*" so likewise Capt. W. W. Hornby (an otherwise successful breeder of this kind) laments his inability to get chickens true to the colour of their parents, and states that this year he had four "spangled hens, but got scarcely any spangled chickens,—and of these half were *double-combed*, though the parents were *single-combed!*" That such should be the case is by no means surprising, when the dashes of true Dorking, Malay, Game, or Spanish blood, so frequently found mingled with the Sussex, together with the constant tendency in all mixed breeds to throw back,—are taken into account: and the only wonder is, that, in the face of the strong presumptive evidence of cross-breeding, which these facts afford, the Sussex fowls should be found carrying off prizes at Exhibitions, side by side with the original and genuine Dorking race; not that we wish to speak in any way disparagingly of the merits of the Sussex *as a fowl,*—for we readily admit that cross-bred birds often surpass their original progenitors; but all we contend for is, that *as a breed* or *variety* they ought not to be permitted to be classed, or to enter into competition, with that of the true-bred white Dorking. In describing the Sussex Fowl we may observe that it very nearly resembles the Dorking in shape, but the body is rather longer and more squat or duck-shaped; it is also a much larger and heavier bird, weighing from 7½ lbs. to 10½ lbs. if a male, and from 7 lbs. to 9 lbs. if a hen; the head should be small, the comb, if single, (which it is more generally than otherwise,) should be large, deeply serrated or vandyked, and of a bright red colour, with large pendulous wattles of the same; bill of a dusky ash colour, or a straw colour; neck longer, more tapering, and not so clumsily carried as that of the Dorking; breast said to be fuller and better " fleshed;" the legs longer than the old breed, and of a grey or slaty shade; the number of claws on each foot perfectly indefinite, varying most provokingly in some specimens from four to five, and even occasionally to six! Those whose birds have only four claws, and to whom marks of true breeding are not objects of interest, contend that the *absence* of the supernumerary claw is one of the great characteristic advantages of the Sussex breed,—whilst the possessors of five-clawed fowls contend warmly that the *presence* of that additional claw proves the purity and genuineness of the stock. We must leave the respective parties to reconcile the two positions,—merely observing in regard to the second, that *although every Dorking Fowl has five claws, it does not at all follow that every five-clawed bird is a Dorking.*

Grey Speckled.—The general plumage of this

sub-variety is a dirty white, indescribably streaked or *speckled* with mixed shades of black, brown, and a sort of dun colour, which are more defined upon the breast and under-parts; the hackle and saddle feathers are straw colour with dark streaks or shadings; the tail usually black, but sometimes broken with white; the legs dusky yellow shade. The hen has much darker feathering, being principally brown speckled with white, black, and yellow ochre.—The *Spangled* are often bred from the speckled fowls, but differ considerably from them in plumage; the breast and under-parts are mottled black and white; the hackle and saddle feathers (the principal distinction) a yellow-tinged white, tipped or spangled with black and brown; back and wings blue black, broken with dark brown and white; the long tail feathers lustrous black, and the shorter feathers white.

Red Speckled or *pied.*—This is a showy kind of fowl, the prevailing colours being dark rich red, speckled or pied with white, and occasional splashes of black; the breast is black and white mottled, slightly streaked with red. The hen of a dark brown or chestnut body, speckled with white.—The *Pencilled* differ from the red pied in the body colours, which are white and black only, and are more regularly and neatly intermixed together; the ruddy tinges on the breast being likewise more apparent.

The *black-breasted* sub-varieties include all the

darker coloured Sussex Fowls; and being rather
handsome and showy fowls, are generally admired
and reared by poultry fanciers. In all probability, they are in many instances the result of the
system of crossing between the Dorking and
Spanish Fowls, introduced many years ago into
Sussex, and which (to the personal knowledge of
the author) was successfully and extensively adopted in that country, and also at Wokingham, in
Berkshire, and in the neighbourhood around.
Their characteristic points unmistakeably betray in
a marked degree the commixture of Spanish blood:
thus we find in the cock, a large upright spiked
comb, pinky skinny face, with a white ear-lobe
large enough to remind us of his foreign ancestor,
whose ash-coloured legs, longer than the true
Dorking, he has also borrowed; the breast, again,
(which slopes rather sharply from the centre bone
to the thighs,) is clothed in the beautiful lustrous
black of the Spaniard, with an ample flowing tail
of the same, with metallic green shades. The
hen, as might be supposed, partakes more of the
maternal Dorking blood, but still possesses a rather
large comb overlapping or falling over on one side
of the head, and a pinky white face, slaty-coloured
legs, and elevated fan-like tail; to which is sometimes added a perfectly black breast. Another
peculiarity is the transition of colour, or rather
discolouration, to which the combs both of the
male and female are at certain periods subject

K

to,—changing from a high red to a sickly pink, and even to a leaden or livid hue,—which every keeper of Spanish fowls will not have failed to observe that breed especially liable to.—The *Black-breasted Silver* (sometimes designated " black-breasted grey ") answers to the description given above ; and is particularly distinguished by the colour of the hackle and saddle feathers, which have a tolerably clear creamy-white shade.—The *Golden* differs from the last only in the substitution of a golden yellow, or deep straw colour, for the white in the feathers of the hackles and saddle ; the wings also being of the same shade, marked, or barred across with black.—The *Japanned* has the entire neck hackles and saddle feathers, with part of the wings, of a rich reddish-brown or copper hue, presenting something of the appearance which has obtained for it a distinctive appellation.

The *Cuckoo Dorking* is so named, it is said, on account of the resemblance of the plumage and markings to the feathers of the cuckoo : its colour is a broken, mealy grey, or bluish dun, barred and streaked with black.   It is very little known and still less admired.

The hens of all the above (with the exception of the Japanned, which is of a deep rich brown, with black streakings) are almost similar in appearance, and are indiscriminately matched with the differently coloured cocks. Among a host of mongrel birds, which even the accommodating Sussex breed itself will not own, we may enumerate the *Black Dorking*, (probably a spoiled Spanish,) the *Muffled* or *Ruffed*, (with an exceedingly ugly tuft or beard,) and the *Hen-cock*, (in which no apparent difference exists between the male and female bird,) all of which might with advantage be consigned to a foundling hospital for nondescript poultry,—or, which would be far preferable, should be exterminated by the cook's knife.

VI. SPANISH. (*Plate* II. *fig.* 3.)

Name.—Spanish ; Le Coq de Caux of Brisson ; Paduan and Jago Fowl, of Dickson and other writers ; Andalusian, Columbian, Ancona, Minorca, and Portugal Fowl. The numerous patronymics need not dismay the amateur fancier of the noble race of fowls correctly and most generally known as " Spanish," there are few other breeds the characteristics of which so completely serve to indicate the personal identity, as do those of the true Spanish, under whatever name they may have to be encountered. There is nothing by which they can be satisfactorily recognised as belonging to the race of Paduan fowls of ancient writers, or of that of Jago of foreign naturalists. The other appellations, though very generally in use in some parts of this country, are simply the names by which different sub-varieties of the Spanish are distinguished, and which will be noticed as such, therefore, in their proper places.

Origin.—In the case of this variety, (if in that of no other,) we think, the popular name given to it may be fairly taken as correctly indicating the source whence it has reached us. It is true it is thought by many to be descended from some of the numerous wild species of the East, first brought from that quarter by the merchants of Spain, and thence into other countries: and there are many things that would seem to render this conjectural origin highly probable. But whatever the source from which it has been originally derived, it is certain that the pure and unmixed race has from such a remote period had its *habitat* in Spain, that, if even it was once a stranger, it is now entitled to be regarded as a native of the soil. It is to be met with in various parts of eastern and northern Europe, and abounds every where along the shores of the Mediterranean,—indeed it is believed, that to the Mediterranean traders of former days we owe this importation direct from Spain: for although the local names of *Minorca* and *Ancona*, applied to some of the degenerated sub-varieties, might induce the belief that those places are the original spots on which it was found,—still, the former, distant only 120 miles from the Spanish coast, may in the first instance have received it from that country, with which a constant traffic is carried on; whilst the latter may also have obtained it (though indirectly) from the same source, it having been, perhaps, conveyed thither by the Dutch, who, from an early period, have traded there. In Holland the breed have long been naturalized: and, within the present century, the best specimens of Spanish Fowls were procured from the Dutch; but we learn from Mr. Baily, that of late that market has been nearly exhausted by the extent of our own demand.

Description.—This fowl is of large size, but of sleek and elegant form, " stately " in its bearing, and remarkably upright in carriage,—the head, when standing erect, being in a line exactly perpendicular with the legs; weight from 5 lbs. to 6 lbs., and the hen about 1 lb. or 1½ lb. lighter: the head is of moderate size only, but adorned with an enormous characteristic, firm, upright comb, of florid red, deeply spiked or " vandyked " on the upper ridge; the comb should rise from near the nostril and extend backwards far beyond the back of the head itself, to which it is *attached* only from the base along the crown, at which point the comb slopes upwards behind; the wattles fine, thin, and long, hanging down generally in a fold like a piece of drapery; the whole face, from the eye downward almost as low as the wattles depend, covered with a large patch of a fleshy substance, something in texture not unlike the wattle, but of a clear white colour, though sometimes marbled or veined with delicate streaks of pink and blue, giving it a livid hue, resembling some of the phases of colour seen in the same kind of appendage in the turkey; this white patch cannot, for high breeding, be too *large* or too *white,* as it forms, together with the large comb, the *principal characteristic* of the true Spanish Fowl, and as such is especially regarded by fanciers; a scarcely less peculiar point being the uniform, *unmixed,* lustrous black colour of the general plumage, with greenish metallic reflections upon the neck, wings, and tail feathers: bill rather long and of an olive shade; neck of moderate length and very erect: the legs should not be too long either in the thigh or shank, although the number of long-legged, mis-shapen cross-breds, which abound in the vicinity of the metropolis, have led many to suppose (erroneously) that *long* legs are characteristic of the Spanish: we have heard some fanciers inveigh against moderate length of leg, as giving the fowl what they term a " duck-legged " appearance,—but the fact is, owing to the want of care in breeding, the Spanish is far too high generally upon the leg, and capable of much improvement in that particular: the colour of the shank should be a clear ash-grey or lavender shade, with the under-parts of the feet of a light colour: some persons, who are well acquainted with this breed, it is true, contend for legs of a light colour, but our own observation induces us to regard the ash colour as one of the characteristics (a minor one it may be) of the Spanish, when true bred: the tail is abundant, flowing, and graceful, with well-arched sickle feathers.

The hen differs' slightly from the cock in form, being smaller and not so broad; also carrying the neck and breast much more forward; the back being at the same time much straighter: the plumage is perfectly black, but not of the lustrous hue of the male: comb large and " vandyked," though not so deeply, of the same brilliant red; it is fine and thin, and from this cause, combined with its large size, it cannot maintain an erect position, but, when attaining the full growth, droops or over-laps *on one side,*—we lay some little stress on the position, because hens sometimes, either from disease or the limpsy texture of the comb, are liable to have it hang over in front upon their foreheads,—a feature certainly by no means desirable: wattles most unfemininely large: the cheek-patch and ear-lobe (though on a reduced scale) likewise form a prominent and characteristic point in this variety; and the colour, as in the cock's face, is most esteemed when of a clear dead white: the fan-shaped tail, composed of nearly straight feathers, is carried almost at right angles with the back and body. The sketches are from prize fowls, the property of Mr. J. Barber, Southtown.

Sub-varieties.—It is a question as to how far it may be right to permit a division in the family of white-faced black fowls, (generally admitted to be the type of *the* ancient and genuine Hispanic race,) by including in our descriptions birds whose points of difference induce a belief that they result from crossing, degeneracy, or even accident, but which nevertheless have come to be regarded as distinct sub-varieties : but following the course adopted with the Dorking Fowl, we prefer bringing them under the notice of our readers, leaving the inquiry to be elucidated further by experience and practical experiments.

Andalusian.—No other member of the Spanish family has a better title than this to be viewed as a distinct sub-variety. The name has of late been appropriated to the recently-imported grey and speckled Spanish : it is proposed, however, to confine it to a *perfectly black* sub-variety, long previously known in this country, and which has been imported from the same district of Andalusia as the greys. In size it surpasses the old white-faced Spanish, weighing, perhaps, 1 lb. heavier: the form seems exactly similar, as likewise does the plumage; the comb and wattles in full-grown fowls attain an extraordinary size; but the face is almost 'entirely covered with a crimson mask or patch,—no vestige of white being perceptible, except occasionally the ear-lobe, which very rarely exchanges the bright red for white: in every other particular, it resembles the true old Spanish. The hen also is of larger dimensions, but scarcely of so graceful a form as the original race, being flatter on the back and not so broad behind; the comb too, though large and over-lapping, is somewhat smaller: in her case, as in that of her consort, we find the white face supplanted by a brilliant red mask.—The *Minorca* fowl of North Devon and Cornwall needs no description, as it very closely follows the above sketch of the Andalusian in every point, though mingled with them we frequently find common black fowls, meriting little more than the names of mongrels.—The appellation of *Ancona* is also very frequently applied to fowls of the Andalusian or Minorca sub-variety,—and occasionally to birds of a mottled or speckled plumage, in which the prevailing colours are black and white; whilst some in addition have the hackle feathers streaked with a reddish brown. With regard to the white face of the original Spanish, some writers have raised the question whether it is not simply the result of very high breeding and domestication; and certainly the fact that, whenever they are bred in for any length of time, the white face gradually disappears, would seem to point to an answer in the affirmative: and if this view is correct, the Andalusian would in reality be the *primitive* breed of Spanish. Our portraits (*plate* II. *fig.* 2) are from birds belonging to Mr. J. Dowsing, of Great Yarmouth.

White and *Grey.*—Instances of these colours occurring in the Spanish are so few that there seems good reason to regard them merely as the result of accidental breeding. They are however usually considered as a distinct sub-variety, and have been styled " Andalusian," a name we have already assigned to the black. Why the lighter plumaged fowls should be called so, it is difficult to say; for Mr. J. Taylor, jun., of Cressy Park, Shepherd's Bush, (who is now almost the sole possessor in this country of birds of that colour,) distinctly informs us that his stock was obtained by a friend in the service of the Oriental Mail Company, who, " after many trials," and with great difficulty, secured 12 fowls of various shades of those colours, " *by scouring the country* (Spain) *far and wide:*" from which it is evident that their title to the name of " Andalusian " rests on very slender grounds. A single pen of the *White,* belonging to this gentleman, were exhibited at Birmingham in 1852; they are identical in form, &c. with the black Spanish, but the plumage is perfectly white: and although most persons who have seen them appear to concur in declaring that the want of contrast and relief in the colouring makes them " *look poor and sickly,*" we confess we cannot comprehend the affected taste that regards white plumage as an acquisition in the Cochin China, and yet repudiates it in another variety. The first white Spanish were imported in 1846.—The Greys include various shades, from a grisly or streaky white and black, to a slaty or smoky dun colour, and also spotted or speckled. The most admired have a slaty-blue plumage with bright black hackles on the neck : the feathers on the breast and body are edged with dark shadings, or marginated, *but not speckled ;* the white face and other points of the black Spanish are present. This sub-variety may, like the white, be the result of accidental or cross breeding, there being nothing indicating them to possess any permanence of character,—as Mr. Taylor admits that the lighter coloured breeds have "become very scarce" in their own country, and also that in breeding, "even after careful selection," they sometimes "throw speckled chicks,"—thus failing to perpetuate their own plumage.

In addition to the above, some few specimens of speckled fowls (black and white) with white legs have been imported from Spain ; they have a large comb, and small white face, but they carry a tuft or top-knot of feathers on the head, with a ruff or beard upon the throat, like the Polands,—which they otherwise much resemble : also some with Spanish characteristics, but of brown plumage, and having the plumes of each feather divided, or double-webbed, so as to give it the appearance of the silky fowl. Sub-varieties of the Spanish are also spoken of having *double* combs (!) with a "tassel" of feathers at the back of the head; likewise one of black plumage with a white, streaky, or mottled breast. All these, though undoubtedly possessing Hispanic blood, might safely be classed together in one lot labelled "*mongrels*," a description to be found every where, and which, in the present rage for importing new breeds, are exceedingly likely to attract the notice of fanciers, their only merit consisting in the heterogeneous nature of their composition, which whilst combining characteristics of the most widely differing races of fowls, yet still is unlike and differs from them all.

VII. POLAND OR CRESTED. (*Plate* IV. *fig.* 3.)

Name.—Poland, Polander, or Polish Fowl; Gallus Cristatus of Temminck; Coq. huppé of Buffon; Paduan Fowl of Aldrovandi; St. Jago Fowl of some writers; and Copple-crowns (a provincialism). The first or English name by which this variety is now known, has proved a perfect riddle, which poultry writers have taxed their ingenuity in vain to solve. Why are they called " Poland ?" All agree that they have nothing whatever to do with that country, being not even known there : and besides, if derived thence, assuredly they would most reasonably have been named *Polish.* Dickson suggests that Poland may have been applied to fowls of this race, on account of some resemblance he fancies exists between the crest of feathers and the square-headed cap of the Polish lancers ! a reason sufficiently far-fetched.

Scarcely less so is that of the Rev. E. S. Dixon, who conjectures the name to be derived from the shape of the bird's skull being like that of a head or "*poll;*" or from the *Plica Polonica*, or *Polish* disease, in which the human hair is matted together—giving a totally different appearance to the crest of a Poland Fowl: others derive it from Polein, the obsolete term for the ancient sharp peaked or " piked " shoes. It is, however, needless to go through the various fanciful and improbable solutions which have been proposed. We generally find that the names of our domestic fowls indicate, either directly or indirectly, if not the actual place of their origin, at least the source whence immediately derived : in this view we will endeavour to suggest some more probable explanation of its cognomen. It is only recently—since, in fact, that they have been imported from the continent—the name of " Poland " has been given them, as they were previously known as " Crested Fowls ; " and it therefore seems by no means unlikely that the term is simply a French designation Anglicised : thus, supposing the generally received opinion that they are of Indian origin be correct, our French neighbours may fairly have styled them *Poule Indienne*, at first perhaps contracted to *Poule Inde* or *Ind'*, whence the transition to the English corruption Poland—almost identical in sound—is easy.

Origin.—This variety, it is thought, is the St.

Jago or Paduan Fowl of old writers; and was introduced by the Spaniards from St. Jago, a town of one of the West Indian Islands, and have thence been widely scattered over the European continent, being domesticated in Holland, France, Belgium, Germany, &c.; they have also been located in various parts of South America, whither they have been conveyed, however, from the Old World; but it is somewhat remarkable that we have no account of the Spaniards (to whom we owe the breed) being possessed of it at the present day. Sonini states that this, or, at least, a crested variety, is extensively reared and highly esteemed in Upper Egypt, and also at the Cape of Good Hope. In this country we formerly obtained the Polands from Germany and Holland: more than a century ago a writer (Albin) gives a portrait of " a peculiar breed," having a top-crest and beard of feathers; which he says was brought " from Hamburgh," and from this cause may have arisen the confounding of it with the Hamburgh variety to be afterwards described. Richardson, who gleaned much of his information from the pages of the earlier editions of our treatise, and who, like many others, may hope to conceal the fact by sneering at the source whence they derive it, takes upon himself to correct an error that was never made: he asserts that the author " errs in supposing the original country of the Polands to have been Holland," whereas Moubray merely states that they " were chiefly *imported* from Holland,"—and such was undeniably the case until within the last few years, when they have been more largely brought from France and Belgium than they were formerly, and from those latter countries our supply still continues to be mainly drawn.

Description.—This is a full-sized variety, though by no means one of the largest. The form is plump and deep, and, if not quite elegant, at least is neat and well-proportioned: the carriage is remarkably upright, so much as to give a very prominent or " pouting" appearance to the breast, and a brisk, sprightly air to both sexes, something akin to the aspect of the little Bantam breeds: the back should be level, though sloping from the base of the neck to the tail: weight, male from $4\frac{1}{2}$ lbs. to $5\frac{1}{2}$ lbs., and the female about 1 lb. lighter usually. The general plumage is of various colours—white, black, or yellow, either singly or in combination; the feathers are exceedingly fine, thin, and soft, giving the bird the appearance of not being so thickly clad as other fowls. The head is singularly round on the top of the skull, and thence springs a very thick tuft, or crest, of long thin feathers or plumes, which falls or " sprays " over and all round the head, covering it as low or lower than the eyes, much in the shape of an umbrella, and giving the face a comical and lack-a-daisical expression: comb, it may be said, there is none,—unless, indeed, the apological manifestation of a very small fleshy protuberance, about the size of a large horse-bean, situated just over the nostrils at the base of the bill, can be regarded as such: this is sometimes seen with two or more diminutive spikes upon the top, but quite as often in two small flattened lobes or halves; the smaller the development of this excrescence the better we should like it, indeed the total absence would not be a matter for much regret. Wattles rather small, (except in the black sub-variety, in which those appendages are of full size,) but on either side, from the cheek downwards beneath the throat, a quantity of pointed ruff-like feathers depend, and, meeting upon the throat, form what is termed a ruff, muff, or beard, more or less prominent in all *true-bred* fowls of this breed,—the blacks again differing in this particular, having no trace whatever of the "beard." In stating our own opinion, that the beard is quite as indispensable a mark of true breeding as the crest or "top-knot," we would guard against this requirement being insisted on too arbitrarily; for many whose judgment is entitled to respect take a totally different view on this point, regarding the *absence* of a beard as necessary as we do its *presence*, to indicate purity, —the Rev. E. S. Dixon, indeed, goes so far as to call the beard "a frightful appendage," and Mr. Baily " never treated bearded Polands as true birds." On the other hand, Mr. Vivian, Mr. Baker, and Dr. Horner, and other eminent breeders of the Poland, strenuously advocate the genuineness and originality of the *bearded* fowls alone: the latter gentleman (to whom we are indebted for the introduction of an undoubtedly very superior strain of this variety) states,—" I have recently had Polands from the continent, Ireland, and other parts, and in all cases *the beardless have been miserably inferior* to the bearded. . . . . . Nay, so evident was this inferiority, that I could readily select the bearded from the unbearded, even when the throat was not visible. . . . . It is highly probable that the beardless Poland is a crossed and mongrel bird, produced between the Poland and spangled Hamburgh fowl. A thorough-bred bearded Poland ever produces *bearded* progeny. On looking over chickens from beardless birds, I find not a few have the double or rose comb of the Hamburgh." The neck is full and strong, amply clothed in a long and handsome hackle; and it is especially in this point, as well as in a deficiency of the crest feathers, that the beardless Polands strikingly betray their inferiority. Saddle hackle consisting of long and fine feathers: tail full and ample, of fan-like form, and carried in a very erect or elevated position, with long sickle feathers drooping above and by the sides: legs of a slaty colour, with lilac tinges; not too long in the shank, nor too thick in the thigh, but altogether neatly made.

The hen is very similar in shape to the cock, and like him has face and throat decorated with (we suppose, in compliment to the sex, it must not be called a " beard ") the like ruff of feathers ; and bears the same tufted crest upon her crown : she appears very short in body, which may perhaps be owing to the elevation of the tail diminishing the length of back from the bottom of the neck to the rump.

Sub-varieties.—The only recognised sub-varieties in the poultry yards of this country are as under :

WHITE (pure colour).	BLACK with white crests.
SILVER-SPANGLED.	CUCKOO, GREY, or
YELLOW or GOLDEN-	SPECKLED.
SPANGLED.	

The *White* fowls with *black* crests were first spoken of by Buffon, as being reared by breeders in his time ; but we are not aware of any authority on which to rest a belief that they were ever introduced into this country,—which, however, it has invariably been assumed they have been, and therefore, being no longer met with, their loss is often deplored by poultry fanciers : certainly, if ever known here, the race has long since become extinct. The wholly white breed we at present possess is limited to the possession of two or three fanciers, and is reckoned scarce. The plumage should be of clear, unmixed colour throughout, with a beard of the same; the crest consists also of white feathers, only partially shaded or mottled with black. A few pens of this breed were exhibited at the Cheltenham Poultry Show, in 1853; and some specimens are said to have been imported lately from the South of France. Mr. W. G. Vivian, of Singleton, is at present the principal breeder of this kind; though some fine birds, with crescent or horn-shaped combs, were lately in the possession of Mr. Tucker, the secretary to the Holmfirth Poultry Association. We have never had an opportunity of inspecting the White Polands, but should have considerable doubts as to their genuineness or originality, and would caution our readers not to purchase these fowls as being " black-crested," a title under which they are frequently advertised: at the same time we are bound to state, that they are said to have thrown, for several generations, chickens of a similar colour to their own; and must admit that they are a very superior breed to the mongrel beardless whites frequently shown. American writers assert that there are in that country some " beautiful specimens " of the true black-crested White Polands— *we should like to see them.*

The *Spangled* Polands when *true-bred* are exceedingly handsome and rare poultry. They are not, perhaps, much sought after, simply because they are not much known—if at all: and it is more than questionable whether, previous to the recent introduction of this sub-variety by Dr. Horner, of Hull, there were any really genuine specimens in this country. That gentleman has taken infinite pains to import nothing but pure and genuine fowls, and the superiority of his stock (which have been " bred to a feather ") is very apparent, both as regards shape and plumage.— The *Silver-spangled* has for the ground colour of its feathers, a fine, clear, transparent white, laced or " spangled " over (the body more deeply so) with markings of a lustrous greenish-black: neck and saddle hackles very fine and long, of a creamy white shade, but some delicately spangled or shaded; as also are the wings and coverts neatly and regularly, but varying a little in the markings in some fowls: under and hinder parts much more deeply shaded—almost approaching entire black: tail full and well arched, black feathers, with the shorter side sickles barred across with white, in half-mourning style: tufted crest large, of dark or black colour edged with white, preserving an uniformity of appearance with the body plumage,—differing in some instances, perhaps, *but always black and white:* the ruff or beard, when perfectly marked, barred black and white. The hens, as ornamental or show birds, far surpass the males, which more usually than not lose the distinct spangles, and become, after the first or second moult, what would be more aptly described as white and black fowls; but the female preserves very tenaciously the dark spanglings, which in her are smaller, but much deeper in shade, more uniform and regular, and extend over the whole body. It may be observed, that in both sexes it sometimes happens that the neck hackles come out at first white, but, after a moulting or two, turn dark edged with white. "Speaking generally," says Dr. Horner, " it is a good fault (if fault it can be called) in Polands that they should be *dark*; as bad birds are those that get without colour, or run into a kind of broken spotted white."

The *Gold-spangled* (sometimes called Chamois) Polands are very handsome and showy birds, differing from the Silver in the *ground colours* of the feathers only,—the greenish-black markings or spangles being identically the same in both sub-varieties. In this, therefore, the crest is black with a deep edging of dark reddish brown; and although few birds have them exactly similar,—some being red mottled with white, and others almost entirely black,—we regard both as imperfections; neither should we care to seek to produce entirely white crests, (which by the way, though said to be "much admired," we never heard of any one having seen,) as it is generally admitted that, in the Spangled Polands, the more the colours of the top-knot harmonize with those of the general plumage, the nearer the bird approaches the standard established in the fancy: the body feathers a rich golden hue, with yellow ochre and rich brown, spangled black: neck and saddle hackles very long and bright golden red: throat "ruffed" with a full black beard.—The hen, as with the Silver, have the spangles much clearer, more regular, and better defined than the male bird; and partakes of a darker feathering, the dark brown being more generally present in greater proportion.—Of the *Yellow-spangled*, stated to be a "new sub-variety," there are far too many specimens everywhere met with, being only an inferior class of the Golden, in which the plumage has lost nearly all the clear and brilliant colouring, the markings less regular, and the spangles dwindled into black fringings or edgings at the ends of the feathers.

Among Spangled Polands the Silver appear to be the more scarce: we learn that Dr. Horner, after much trouble and searching out personally by friends, obtained several Golden Polands from the continent, but not a single Silver one.

The *White-crested Black* sub-variety, though far more generally known than any other, differs in many points from the Spangled: and if, as has been conjectured, the Polands have some distant connexion with the Spanish race, in the Black fowls the presence of Hispanic blood may perhaps be suspected. The form is rather larger and rounder than that of the Silver or Golden, and scarcely so neatly shaped: the plumage is entirely a lustrous greenish black: the head surmounted with a large and abundant tufted crest of feathers, perfectly white, except at the roots, on the forehead, which are black; on this point Mr. Baily gives it as his opinion, that there should be no black feathers whatever in the crest, but he at the same time says, he "*never yet saw any without.*" We believe it is an established fact, that birds of this breed are invariably without ruffs or beards; but have long scarlet wattles depending from the cheek; white ear-lobe; with dark-coloured bill and legs. The tail should be a pure unmixed black; and, indeed, any white in the body plumage, except some light downy feathers beneath the tail, is looked upon as a decided blemish. The hen has a more compact and rounded crest, the feathers being rather shorter and closer than in the male: she has, however, only very small wattles. In both a very little comb, or rather the germ of one, is almost always present. Some years ago, the Polands imported were uniformly of this Black sub-variety, which thus gave rise to an erroneous impression that the Spangled fowls (which were long previously known and reared upon the continent) were produced from crossing with other varieties in this country.

Our portraits are taken from prize birds, the property of Mr. James Barber, of Great Yarmouth. (*See plate* IV. *fig.* 1.)

The *Grey, Speckled,* and *Cuckoo* fowls of this class, comprise a motley group, with only very slight claim to the title of Polands. Of these the uniform *slaty-blue,* or dun-coloured, have perhaps the best title to be treated as distinct, although probably from some foreign breed. The *Grey, silver pheasant,* or black and white spotted and speckled, sufficiently indicate by their appellations the description of their plumage,—the former having dark grizzled feathers, whilst the latter have various shades of white and grey, broken or spotted with irregular patches of black, having also a blue cheek-patch and small double comb. The hen is much smaller, and is more regularly marked or speckled: it is called the " Barbary Fowl " by Richardson. The *Cuckoo,* like other fowls bearing that prefix to their names, having plumage resembling the feathering of a cuckoo's breast. It is very questionable whether any of the above are genuine sub-varieties.

VIII. HAMBURGH.

Name.—Hamburgh or Dutch Fowl ; Gallina Turcica,(or Turkish Fowl,) of Aldrovandi ; Everyday and everlasting layers ; Leghorn Fowl of American writers ; and Red Caps (provincialism).

Origin.—This highly ornamental and not less useful variety, is perhaps another breed of Eastern origin,—the name of Turkish Fowl inducing a belief that they may, possibly, have been originally transmitted to Europe by that way: whilst they were introduced into this country from Hamburgh and Holland, where they have long been located, having been no doubt imported by the Levant traders. Wherever may have been their primitive *habitat*, it is tolerably certain that our own stock has been derived from Germany and the Netherlands. Although long known and reared in Lancashire, Yorkshire, and the north of England, (where they continue to be held in great estimation,) they were first brought into general notice through the medium of the pages of this treatise, in a communication to the author, from the Rev. J. A. Ashworth, vicar of Tamworth, near Bolton; previously to which they were quite unknown in the vicinity of the metropolis, and indeed in most parts of this country, if we except the above-named districts.

Description.—The size of this variety is small, certainly of smaller proportions than the Game, which, however, for neatness and symmetry of form, it quite equals,—the contour being sleek and graceful: the weight averaging about 4 lbs., and never exceeding 5 lbs., in the male. The head is small and flat, and, *when pure-blooded*, invariably surmounted by a fine crimson double or rose comb, consisting of a broad, flattened fleshy mass, serrated upon the upper surface with a number of small spikes, so as to present the peculiar quilled or frilled appearance that has obtained for it the term of a *rose* comb; this comb is smaller in the Silver than in the Golden sub-varieties;—but both are distinguished by a peculiar and characteristic elevation of the comb at the termination on the back of the head, which is a single point or peak, always found growing upwards as it were: wattles long, large, and folded; with pinky face, and full white ear-lobe or patch: eyes large, bright, and unusually prominent: the breast (more particularly in birds of the Golden class) large and much developed as compared with the rest of the body: general plumage what is called "close-feathered:" neck and saddle hackles thin, but very long—especially the latter, which hang down far below the rump: wings small, but strongly set on: legs short, slender, and of light ashy or bluish grey shades: tail much elevated, well arched, ample, and with long, graceful, and waving plumes.

The hen is very little smaller than the cock; and has a fine red rose comb, though smaller and flatter; pinky face, white ear-lobe, and wattles much larger than is generally found in the females in poultry. The form is very similar to that of the cock, but the body seems rather longer shaped; the plumage is always more strongly and distinctly marked; and the under-parts are furnished with some fine flossy feathering: tail long, flat, and tolerably elevated.

Sub-varieties. — The multiplicity of different

names by which the same sub-varieties of Hamburgh fowls have hitherto been designated, rendered their classification a perplexing matter; and although the arrangement proposed by the committee of the Birmingham Exhibition, clears away much of the confusion in the nomenclature, we think the plan of dividing them by the specific *markings* of *similar colours*, scarcely so simple or correct as it might be rendered by separating the race into two divisions,—including under one, all the *white* or *silver* feathered, and under the other, all the *yellow* or *golden* plumaged birds; these again being sub-divided according as the variations in the markings are pencillings or spangles. In the following arrangement it has been attempted to present a complete list of the local or provincial names given to the four sub-varieties in which the Hamburgh family is comprised.

WHITE.

Silver Pencilled,—or Bolton Greys, Cheteprats, Chittiprats, Corals, Creels, Creoles, Kriels, Narrowers, Pencilled Dutch, or Prince Albert's Breed.

Silver Spangled,—or Silver Pheasant, Silver Mooneys, or Silver Moss Fowls.

YELLOW OR RED.

Golden Pencilled,—or Bolton Bays.

Golden Spangled,—or Golden Pheasant, Golden Mooneys, or Copper Moss Fowls.

The *Silver Pencilled,* or Bolton Greys, have long been esteemed by many as the most valuable and handsome sub-variety, though they are certainly not so showy as the Golden with the same markings, and are rather more diminutive. The ground colour of the feathers is a clear silvery white—*invariably pure and unbroken upon the lappel or neck hackle;* and upon the rest of the body more or less pencilled or marked with a bright greenish black,—each feather being barred across the webs with five or six of such pencillings; and these points constitute the principal characteristic distinction between the *pencilled* and *spangled* varieties,—the difference in the appearance of which is not otherwise so easily distinguished. The male in this, as indeed in the other sub-varieties, has not the pencillings so clearly displayed as the female, —the back, breast, and saddle hackle, as well as the neck, being unbroken white, and having only the wings and *sickle feathers* of the tail slightly marked with black—and even these sometimes disappear with age: the tail itself entirely of a lustrous black colour, with the exception already made. The hen is marked all over very regularly and clearly, contrasting prettily with the pure white of the neck hackle; each of the tail feathers barred to the very tip.

Most of the provincial names applied to this breed, would seem to have reference generally to their diminutive proportions: thus we have the Yorkshire *Chittiprat,* denoting a " starveling," or " puny brat;" whilst the *Corals, Creoles,* and

Creels, of the south, though supposed by some to have their origin, the first from the red or coral-like comb, and the second from the mixed black and white plumage, are, we think evidently merely corrupted forms of the old word Kriel, formerly used to express a pigmy, or diminutive race of beings.

Silver Spangled.—This is of decidedly larger proportions than the preceding; but otherwise it much resembles it in the general colours and appearance of the plumage,—although the markings or *spangles* are sufficiently distinct from the pencillings. The ground colour is white, the *ends only* of each feather being tipped with a single circular patch of black, not quite so large as a sixpence: these spangles, or *moons*, (hence the term Mooneys,) in true-feathered birds, are formed perfectly whole and clearly defined, and when thus, the feathers being neatly and regularly placed or *tiled* over each other, the spangling is of a remarkably uniform character, which no other breed exhibits in such a degree: any indication of the spanglings running together, or " clouding," is a most incurable defect on any part of the fowl. The hackles are of creamy white, the centre of each being streaked down with black: *tail itself* spangled throughout; but the *sickle feathers* greenish black, sometimes slightly barred or streaked with white.

There is not a very great difference in size

between the male and female; but the latter has a much smaller comb, is more distinctly and regularly marked; her tail, too, (which is in the shape of half a lady's fan,) is quite white, except just at the ends, where the feathers are tipped or spangled with black.

Golden Pencilled.—This is a more showy and uncommon breed than the Silver; and it is also rather less diminutive in size. The ground colour of the plumage is of a rich golden hue, varying from a light orange to a deep orange red, and reddish ochre,—the dark pencillings upon the feathers being identical in colour and arrangement with those of the Silver Pencilled, the description of one sub-variety serving for the other, if we only substitute the *ground* colour. The Golden has a full rose comb and long wattles; a fine large chest; neck and saddle hackles long and flowing, of a particularly brilliant golden red; and tail a metallic black, with green reflections prevailing.— The hen has a neck hackle of the same hue as the male, but the rest of the body is more of a gamboge or ochre, and darker upon the wings, coverts, and tail.

Our portraits (*plate* IV. *fig.* 2) are those of prize fowls, the property of Mr. Richard Steward, Great Yarmouth.

In Lancashire the pencilled Hamburghs are known as " Bolton Greys," and " Bolton Bays," (accordingly as they are silver or golden,) taking that name from the town where they were originally most extensively reared.

The *Golden Spangled* has the same colours exactly as the preceding, but the dark markings consist of only *one* black tip or *spangle* on each feather, similar to the Silver Spangled fowls, like which also they are often locally called " Mooneys."

The *Moss* fowls are of two kinds, silver and copper, and are generally classed among the spangled sub-varieties ; but appear to us to range *between* the pencilled and spangled, (of one or other of which, it has been suggested, they are imperfectly feathered specimens,) the plumage partaking of a *speckled* or splashed character, giving it the broken appearance that has obtained for these birds the name of " Moss " and " Pheasant."   They present the general characteristics of the Hamburgh, but have larger combs and wattles, and smaller white ear-lobes.   In the Silver Moss the hackles, back, and hinder parts are white, streaked or speckled with black irregularly : the wings have more white in the quill feathers, and more black on the coverts : breast and under-parts clouded with black marks or patches : and the tail black splashed with white.   The Copper Moss answers to this description, only substituting a reddish or copper hue in those parts of the plumage which in the Silver Moss are white ; the neck hackle, also, being of ochre, brown, and greenish-black shades, intermingled : the tail of the latter shade, but mottled with ochre and white near the roots. In both these sub-varieties the breast feathers are very often deeply clouded with black, and are even sometimes found entirely black,—although this is more especially the case in the Copper Moss. They are highly prized in some parts of the north, but are more extensively popular in the midland counties.

IX. SCOTCH BAKIES.

Name.—Scotch Bakies or Stumpies; and called in England Barkies (erroneously), and, more aptly, Dumpies. The names, which have all evidently a reference to the short legs, and squat, lumpish appearance, afford a striking illustration of the facility and rapidity with which the transformation in the nomenclature of poultry is effected.

Origin.—This novel and highly valuable variety (whose many useful qualities will be hereafter noticed) was introduced by Mr. John Fairlie, of Cheveley Park, near Newmarket, in the year 1852, from Scotland, where they have long been known, and are great favourites, more especially with the poorer classes. From a communication with which Mr. Fairlie has most obligingly furnished us, and from which we have almost exclusively gleaned the subjoined particulars,—it appears by no means unlikely that, however distinct they may be now, the original founders of their race were Dorkings, perhaps stunted or dwarfed by the coldness of northern climes; and this view is rendered the more probable from the fact that many of them are found in the possession of *five claws* to their feet.

Description.—This variety closely resembles the Dorkings in form, shape, and general appearance; the body is round, plump, and deep, but not so large,—the average weight of a Bakie cock being about 6 lbs., reaching sometimes an extreme weight, with high feeding, of 7 lbs.; the hen a pound lighter: head rather large and surmounted by a full single comb of deep red, erect and well serrated on the upper ridge; wattles long, thin, and of pinky red shade, with cheek-patch of the same, and full white ear-lobe; neck strong and thick, and long in comparison with the *stature* of the bird; chest remarkably fine, broad, and fleshed; the general colour of the plumage is white, tinged with light lemon or yellow,—which, indeed, nearly all white varieties more or less exhibit; fine and full hackle feathers; back of head and neck is more deeply shaded with a greenish lemon hue than any other part; breast, saddle, and underparts very nearly the colour of the breast plumage of the clouded or speckled Dorking; wing coverts pure white; tail full plumed, and flowing sickle feathers, colour most admired when clear unbroken white,—though sometimes found slightly streaked with black; the legs, which form the principal distinctive feature of the breed, are

M

extraordinarily short, "never," says Mr. Fairlie, " exceeding 2 inches from the hock joint," from which circumstance they seem to be sitting or squatting even when standing. The same gentleman assures us that all the young he has reared have proved singularly true to the parent stock— which goes far to prove the distinctive character of the breed. The hen presents no marked points of difference with the male, worthy of note,—except, perhaps, a much smaller comb and wattle, and an almost horizontal tail, scarcely ample enough to be graceful, when viewed on a comparatively large body.

Through the kindness of Mr. Fairlie, (by whom, with Lady Paget, of Sennowe Guist, the breed is exclusively possessed in this country,) we are enabled to give sketches of his prize fowls. (*Plate V. fig.* 3.)

X. BANTAM.

Name.—Bantam; Gallus Pusillus of Temminck; Gallus Banticus of Brisson ; and Coq de Bantam of Buffon. The name is evidently derived from the town so called, on the north-west coast of the island of Java, in the Eastern Seas.

Origin.—The Bankiva has already been referred to (page 70) as the progenitor of our own pigmy race of Bantams ; which it seems tolerably clear must have been conveyed from the town of Bantam, once a flourishing place, but now a miserable village. Perhaps the breed was imported some two centuries and a half ago, when Bantam, the great pepper depôt of Java, was in our possession. Some few years later than the period above referred to, Aldrovandi, the Italian naturalist, described " a pigmy or dwarf sort," of which, in consequence of its " rarity," he gave a rude " figure."

Description.—The form of this Lilliputian inhabitant of our poultry yards, notwithstanding its extremely diminutive proportions, is plump and full; the back is very short, and the breast remarkably prominent; the size may be estimated from the extreme weight to which high-bred specimens are limited,—namely, 18 oz. for the cock, and 15 oz. for the hen. The general appearance is singularly gay, lively, and grotesque— that combination of effect being produced by the mettlesome, proud, arrogant bearing, the conceited, sort of self-satisfied air, and the impudent, coxcomb-like strut, which these little Turks exhibit. The head is small and rounded, and surmounted by a full rose or double comb of deep rosy red hue, as also the cheek-patch, with a small white ear-lobe, and full-size wattle of bright red; fine neck and saddle hackles; wings depressed and projecting downwards about two inches below the body, in a way similar to those of the Malay, but in a greater degree; tail full and elevated, with good sickle feathers, well arched inwards towards the back of the head, which is thus brought in close proximity : legs sinewy, but slender, and generally of ash or lead colour; the shanks of the legs should, we think, always be clean and unfeathered, although some years ago there were many specimens of Bantams (imported, we suspect) in this country, having legs as heavily feathered, or "booted," as it is called, as any of the modern Cochin Chinas; and from this cause it was formerly imagined that true-bred Bantams should *not* have *clean* legs,—though the reverse is now the generally accepted opinion.

Sub-varieties.—It is difficult to say which of the sub-varieties can be regarded as the type of the Bantam race. This breed (like most others) has been split up into numerous sub-families, mainly owing to the fact of the degenerate and mongrel individuals connected with it, from their frequent recurrence, coming to be regarded as independent and distinct; but acting in accordance with the plan we at first started with, of endeavouring to render the enumeration and classification of our domestic poultry as concise and simple as possible,—we shall not hesitate to restrict the division of the family to the following narrow limits :

WHITE.	BLACK.
YELLOW or NANKIN.	SEBRIGHT. { Silver laced.
SPECKLED or PARTRIDGE.	Golden laced. }

White.—This is a very neat and pretty breed,

but having the disadvantage of being rather larger than others of its family—a serious fault in Bantams, which, however, has of late been much obviated by more careful attention to breeding. The plumage is entirely of clear unbroken white colour ; cheek-patch of a rosy hue, with white ear-lobe ; bill and legs are of a creamy white ; comb very fully developed, and of deep and bright red : altogether it forms a pleasing relief and contrast to the other variations of colour in the members of this pigmy race. A short time since, Mr. Baker, of Beaufort House, Chelsea, had some exceedingly beautiful and small specimens of this sub-variety, " booted " with feathers on the legs, almost as long as the legs themselves ; but whether bred from stock similarly accoutred we know not, and assuredly feathered-legged Bantams have almost entirely disappeared in this country.

Yellow or Nankin.—This is an old-fashioned and once popular sub-variety,—although now but little prized as fancy birds. The name has been bestowed upon it from the similarity between the colour of the plumage (especially that of the hen) and the yellow shade of the fabric vulgarly known as " nankeen." The cocks vary much in the feathering, from an orange to a brilliant red and chestnut, with bright metallic hues on the wing coverts, like the duck-winged Game : and from the great resemblance both male and female often exhibit to the Game family generally, many writers have been induced to style them " Game Bantams," a designation we do not think desirable to retain in reference to this sub-variety, as it is calculated to lead to the confounding of the Yellow or Nankin with the questionable race of Bantams (so frequently seen) with presumptive evidences of a mixture with the Game family: the " *black-breasted red* " Bantams, for instance, which have recently been elevated into a distinct sub-variety, though certainly of very small proportions, in every particular are almost identical with the Game fowls of that name,—possessing precisely the same thin *single comb*, wattles, &c.   The Nankin hen is of a paler tint of yellow, sometimes called a " ginger yellow," with a very small comb and wattle, and dull leaden-coloured legs : the feathers on the neck are often streaked, but this is regarded as a defect,—a clear-coloured plumage being the principal mark by which this sub-variety is separated from the succeeding one.

Under the *Speckled* division of the tribe, we should include those Bantams usually known as " Partridge " and " Pheasant."   The plumage of a dark ochre with a rich brown and greenish black ; the neck and saddle feathers being especially marked or streaked down with the latter colour ; the back and wing coverts display that peculiar feathering termed partridge or grouse-coloured ; tail entirely black.   Birds of this sub-variety are often found with more brilliant plumage, and more neatly marked,—approaching closely to that of the Golden Pencilled and Spangled Hamburgh, whose peculiar spiked comb is likewise present, with legs of a similar bluish shade,—indeed the close resemblance in those, induces a suspicion that such specimens may actually be the offspring of some *liaison* between the small Hamburgh and the still more diminutive Bantam breed.

The *Black.*—The rich glossy and lustrous black plumage of this little inhabitant of our yards, is too well known to need a detailed description. The full rose comb, pink face, pure white ear-lobe, and long scarlet wattle, dusky olive-coloured bill and legs, with, in short, all the Bantam characteristics, added to the *smallest possible proportions,* should be rigidly looked for in this sub-variety. Some little prize fowls, formerly belonging to E. H. L. Preston, Esq. of Southtown, furnished the originals for our sketches. (*Plate* V. *fig.* 2.)

Sebright Bantam.—This breed, which bears its name in honour of Sir John S. Sebright, (formerly M. P. for Hertfordshire,) was, we believe, first brought into notice in this country, about the year 1830, since which time they have risen rapidly into favour, (as much perhaps owing to the comparative scarcity, as to anything else,) until at the present time they are regarded as the perfection of Bantam breeding. An idea has long prevailed that Sir John Sebright was the " originator " of this sub-variety, or in other words, that by long and careful breeding and crossing of other known branches of the Bantam family; and by some mystical operation nobody attempts any explanation of,—the worthy baronet effected such a revolution in nature as was never done before or since, and wrought such a change in the existing birds of that kind, both as regarded feather and feature, that the result was the production of an entirely new, distinct, and permanent sub-variety. It is not, however, easy to divine by what means this opinion can have obtained in the public mind; for assuredly, although Sir John may not have chosen to disclose the source whence he procured the laced Bantams, he never put forth any claim to be considered the " originator " of them; it is true that he was the first who introduced them into this country, (wherever obtained,) and that from his yard alone, directly or indirectly, the existing stock (limited as it is) has been descended; but there is nothing on which fairly to found the supposition alluded to above, nor do the notices published at the time this breed made its appearance afford any grounds for it, but on the contrary: thus in one of the early editions of this treatise, which appeared shortly after the Sebright Bantams were first brought out, the author there speaks of the breed as having " been lately *obtained*," and of Sir John himself as " *one of the chief amateurs* " of that sub-variety. Everything, therefore, seems to favour the conjecture that this showy race of Bantams, like the other members, were originally inhabitants of an Eastern clime, imported, perhaps, by Sir John Sebright, and certainly successfully reared by him in the greatest perfection.

Description.—In regard to size, we should concur with Mr. Baily in placing the Sebright among the very smallest of this dwarfish race,—fixing the extreme limits of weight at 2 lbs. for the pair, and as much under 30 oz. as possible; indeed that gentleman states that he has seen an adult pair, perfect in every point, weighing only 23 oz. Independent of the peculiar marking of the plumage, the Sebright cock presents one or two features in which they widely differ from all others of their tribe; instead of the fine luxuriant hackle feathers which cover the neck and saddle in other Bantams, the Sebright is distinguished by the total absence of that long plume-like dress, those parts being clothed in the same short feathers as are found on the rest of the body; and in place of the usual arched or sickle feathers in the tail, we find some broad, straight, and flat feathers only, in the form of half a fan. The comb is double, very broad (comparatively) at the base on the forehead, but tapering gradually backwards to a point; the wattles of moderate size, with rosy cheek-patch, and large ear-lobe of a livid white hue. The tail is carried very high and much elevated—almost at right angles with the back, rendering the resemblance in this particular, and in the general carriage, to the Fan-tail pigeon, more striking than in any other sub-variety.—The plumage of the *Silver-laced* is much admired when of a clear, transparent, bluish-white ground colour, although the shade more frequently met with is that lemon tinge which only too often detracts from the beauty of the white-feathered fowls: all the feathers from the head to the commencement of the tail, and not excepting the wing flights, are very beautifully and distinctly fringed or marginated the whole way round with a black edging of uniform width, giving it the appearance of having a piece of net-work, of small but regular meshes, thrown over it, and which peculiarity has obtained the very apt term of " lacing : " some writers have given it as a law that this black edging should not be wider than the sixteenth of an inch ; but in *young* fowls we should not regard a broader lacing as any defect, on the contrary, it should rather be considered a great acquisition in such cases, for there is a very great tendency in all pencilled or spangled birds (and more especially in the Sebrights) to " *run out of colour*," that is, to lose the dark markings which constitute the chief beauty of the plumage ; a little extra depth in the marking will, therefore, always be expected to soften down as the bird gets older : all we should insist on is, that the lacings be distinct and unbroken, of uniform widths, and placed equi-distant from one another.　The tail should be of the ground colour, not only tipped with black at the ends, but also fringed on the edges all round.—The *Gold-laced* answers exactly to the above description, as regards the markings and arrangement of the colours ; but the ground colour is of a deep rich yellow or ochre, with a brownish hue,—something the shade of very dark amber, but more bright : the dark lacings are precisely the same as in the Silver-laced.

There are few breeds so little known, or which have been so misdescribed, as this,—most poultry writers omitting all description of it, or where any description has been attempted, the errors render it worse than useless: in the very last edition of Richardson's little hand-book, for instance, it is said that the " high-bred " Sebright cock should have " *full hackles*," a " *well-feathered* " tail, wings " *barred with purple, tail feathers and breast black !* " From the peculiarity in the *hen-tail* form of that appendage in the cock bird, the designation of " *Hen-Cock* " has been most improperly applied to it. So, also, they are frequently incorrectly called silver and gold *spangled* instead of laced,—the former phrase denoting a *spot* or *spangle* at the end of the feathers, and therefore conveying a very imperfect idea (or rather, none at all,) of the *edging* or *fringing*, which is so appropriately termed " *lacing*."

NONDESCRIPT AND ANOMALOUS BREEDS.

UNDER these heads we propose to notice all such races of domestic fowls as, although exhibiting many points of resemblance with, and partaking of the characteristics of, existing and distinct varieties, yet nevertheless, owing to some striking peculiarity they possess either in form or feather, cannot properly be classed with the established families, but appear rather to inherit a character *sui generis,* as it were, which baffles equally the naturalist and the student in ornithology. Again, there are many fowls which, by a system of hybridizing long persevered in, seem to have attained a permanent distortion from the natural character of both their original progenitors, and can only be regarded as examples of *perpetuated mongrelism ;* this, indeed, may also result from accidental but frequent intermixture of varieties,—for it must be obviously apparent to every one, that *mixed* or *chance-medley breeding* must be very generally prevalent among poultry, as well as among all classes of domestic animals, where due care and attention to this important particular is not observed. The superior breeding districts have always been few, and in the mean time there have been particular varieties in many other parts which have seldom attracted notice elsewhere, or beyond their immediate locality, and which have been injudiciously permitted to intermix at pleasure, and in the most promiscuous manner possible.

We shall therefore briefly notice the remaining breeds of fowls, which though possessing, perhaps, no quality as domestic poultry worth preserving, can yet scarcely be altogether passed over. We shall attempt to arrange them in the best order their anomalous nature admits of—describing them in the arrangement of the *preceding classes*, accordingly as they seem to us to partake more or less of the predominating character of any particular variety, to which they may be directly or indirectly referable : the class, therefore, under which they are placed in the following descriptions, must be taken simply to indicate the established breeds to which they exhibit *some affinity*, however distant—as Cochin, Malay, Dorking, &c., as it may occur to our judgment is the case.

I. BARN-DOOR OR DUNGHILL FOWL.

Of indescribable shape and form—of unlimited proportions and undefinable size—of every imaginable shade of colours — of unknown parentage and combination of species — without any fixed local habitation and name,—our readers will not be surprised at our excluding it from a place among the *varieties* of domestic fowls, and must excuse us for dismissing this imaginary *species*—

this beau-ideal of the gallinaceous tribe, (according to some writers,) in the same category as that to which we have already consigned many others of heterogeneous character, passed in review in the preceding pages.

II. EMU FOWL, OR SILKY COCHIN. (*Class* I.)

Although this fowl has been named after the Australian native, from some indistinct resemblance which the plumage has to the peculiar woolly or hairy texture of that of the latter ; still, we think, it much more nearly approaches the very singular and characteristic feathering of the Silky Fowls, to be afterwards described. The so-called " Emu Fowl" of poultry fanciers is of large proportions, and otherwise, as regards shape and form, presents great similarity with the Cochin China,—whose meagre supply of tail feathers is, however, yet further abridged in this breed. The colour of the plumage is said to be most usually of a dark buff or fawn colour,—not much unlike some of the lighter parts of the fur on a hare-skin. As only one specimen (and that a female) has, we believe, reached this country at present, little can be said about this fowl,—and, indeed, it may be only a solitary example of the anomalous and eccentric course which nature sometimes exhibits.

III. SHAKE-BAG, OR SHACK-BAG.
(*Class* III. *and* IV.)

This was formerly the largest fowl known, and was among Game cocks unsurpassed for courage and prowess: it is said to have received the singular appellation it was known by, from the custom of cock-fighters taking their birds to the pit *in bags*, which were then *shaken* by way of challenge to the others present,—though we confess to regarding this derivation as much too far-fetched to be correct. They were also called the " Duke of Leeds's breed," for the same reason as Sir John Sebright gives the name to a Bantam variety,—because his Grace (who was a celebrated amateur breeder of Game cocks) was supposed to have originated or introduced the breed some eighty years,—but how or whence is as much a mystery as the Sebright Bantam. The only Shake-bag ever described in this country was one possessed by the author as far back as 1784, and which was purchased from Goff, the noted poulterer, then of Holborn Hill: it was an enormous bird, weighing about 10 lbs., and the colour said to be " red." Its great size enabled the poulterers and inn-keepers of some places, on an emergency, to provide one of these fowls as a convenient substitute in the absence of a turkey. Not very many years after their first introduction, the Shake-bag breed is thought to have become " worn out," and at present to be perfectly extinct, notwithstanding efforts made to preserve them by constant breeding with Malays. But it has been very properly observed, breeds of animals seldom become extinct without leaving some representative traces behind; and Mr. Martin states that, some years since, he possessed " a black-breasted yellow Game cock," of immense proportions, reared in the north of Yorkshire; and numerous victories in severely-fought mains attested its surpassing qualities for the cock-pit. Some short time since, Mr. Robert Pratt, of Great Yarmouth, had some gigantic Game cocks believed to be veritable Shake-bags; and others of a similar description were to be seen in the yard of Miss Sheriffe, near Southwold. Dr. Bennett, of America, describes a pair, " exceedingly rare " in that country, imported [where from?] by Mr. J. F. Tucker of Boston, U. S.; the average weight, the Doctor states, is " from 8 lbs. to 14 lbs., (a tolerable " long range " certainly,) and they are described as of " brilliant plumage," bright red and yellow, beautifully shaded black: but the comb and wattles are stated to be " large and single," whilst the so-called " portraits " that accompany the account, represent the latter short, and the former of a small, stunted Malay form; so which is the more correct we cannot undertake to determine.

III. COLUMBIAN.   (*Class* III. *and* VI.)

The existence of this supposed breed is spoken
to by Richardson, and, from him, by subsequent
writers.   He calls it " a noble fowl," a " native
of Columbia, on the Spanish main, in South America:" though, singularly enough, it seems *quite
unknown to the Americans themselves*.   Richardson speaks only of some specimens the property of
two gentlemen in Ireland; but he does not attempt to give any description of them.   In the
Cottage Gardener they are said to be "larger than
the Spanish, with fine velvetty black plumage,
and tendency to being ruffed or bearded,"—a description that, taken in connexion with the fact,
that, notwithstanding the *furor* for new varieties,
they have not yet found a place at the Exhibitions,
induces us to regard them only as (perhaps inferior) specimens of the black Andalusian Fowl, already described at page 136 *ante*.

IV. RUSSIAN, OR SIBERIAN. (*Class* V. *and* VII.)

This fowl we believe to be nothing more than a
cross between a Dorking and a Poland,—the
northern name having probably been given it,
from the confusion in which one of its progenitors
(the Poland) was involved in the misconceived
idea of its having been imported originally from

the country of the Poles. The Russian is more usually met with of large size, and broad deep build: the cock has a tufted beard, but very inferior to the Poland; as, also, has the hen, which has not unfrequently in addition a few straggling tufted feathers drooping backwards from the head, —but both possess a moderate-sized comb. They are varied in colours,—some a greyish white, some white indistinctly barred or speckled with black patches, and others presenting an admixture of a dark tawny red with brown and black—very similar in plumage to the dark-red speckled Dorkings.

V. LARK-CRESTED. (*Class* VII.)

This fowl, which the Rev. E. S. Dixon dignifies, in a separate chapter, as a perfectly distinct breed, bears unmistakeable marks of being nothing more than a degenerate, ill-bred, or dunghill race of Polands. They are inferior in size and shape to the last named, having a more elongated body, and of more angular contour: the carriage also is far inferior; and it sadly lacks the neat legs of the Poland. Still greater difference is apparent in the crest of the fowl, which in this breed consists of a few long straggling feathers, quite at the back part of the head,—sometimes comprising only about half a dozen in number, whilst the cocks are as often found with the crest entirely absent, the place being supplied by a rather large single comb. Their colours are, as might be supposed from their origin, exceedingly varied—running from a pure white, through white and black speckled or *patched*, rusty black or brown and yellow, (answering to their spangled prototypes,) and so to a perfect black plumage. They have no ruff or beard, as that appendage would seem, by careless breeding for a long time, to have disappeared, and by the same process the crests or top-knots (as they are called in Polands) are getting more and more scanty.

VI. CREVE CŒUR. (*Class* VII.)

This fowl constitutes one of several *raræ aves*, recently introduced, for which the poultry world is indebted to W. G. Vivian, Esq. of Singleton, near Swansea, a very great amateur and most successful breeder of the Poland variety, and in whose hands, we believe, the birds about to be noticed are at present exclusively. The Crêve Cœur is stated to be from Burgundy; and appear from the descriptions given to be bearded Lark-crested fowls: they are of large proportions, have the occasional crescent-formed comb of the Poland, with a scanty apology for the tufted crest and beard. They are either of a mixed gold and black colour of plumage, or else black and white mottled,— "somewhat," says Mr. Vivian himself, "after the fashion of an irregularly spangled Poland fowl."

VII. NORMANDY. (*Class* VII. *and* V.)

This is a long-bodied bird, with a small tuft of feathers sticking bolt upright on the top of the head. An unfortunate additional claw betokens, in connexion with other adjuncts, no very distant connexion with some member of the Dorking tribe, whose speckled black and white attire it is also clothed in.

VIII. BRUGES. (*Class* VII.)

This answers the description of an ill-bred unbearded slaty-blue, or dun-coloured Poland, noticed in our account of that fowl.

IX. BRAZILIAN.

This is a fitting illustration of just such a chance-medley breed as, we could imagine, would be produced, if it were only possible to hatch a chicken from an artificial *ovum*, which should contain the yolks of the eggs of the Malay, Game, Spanish, and Poland varieties.

X. BAVARIAN CRESTED. (*Class* VII.)

This bird, which seems to be circumscribed to the American poultry yards, is described as having black plumage and legs, (the latter often " heavily feathered,") with a crest and " whiskers of mottled black and white ; " and, usually, a large ample tail and long wings.

XI. BREDA. (*Class* VII.)

An importation (we presume) from the Netherlands. It is of full size, perfectly black plumage, and possesses feathered or " booted " legs; and but for the great singularity that it has neither comb nor crest of any kind, it would very well answer the description of the Bavarian Crested Fowl described above. But perhaps, making allowance for the heightened colouring given to their fowls by our transatlantic brethren, the Breda Fowl may be identical with the " Guelderland Fowl," caricatured in the *American Poultry Book*, page 81, imported some years since at Boston, (U. S.,) from the north of Holland, where they are said to have "originated." They are described of blue-black plumage, heavily-feathered black legs, having no comb, but "a small, indented, hard, bony substance, instead," at the back of which " rises a small spike of feathers." The striking peculiarity which the Breda Fowl presents, makes us regret the ill success which has attended Mr. Vivian in his efforts to secure the perpetuation of his stock,—an attempt not likely to succeed, as he complains of the extraordinary number of *infertile* or *addled eggs* which they produce.

XII. JERUSALEM. (*Class* VII. *and* III.)

The stock of this breed was obtained by Mr. Vivian, some six years ago, from the Holy Land. They possess many points indicating them to be the result of a cross with the Malay and Silver-spangled Poland. The ground colour is a dead white, with a light yellow or lemon tinge, and the hackle feathers are distinctly striated or marked with black; which colour is also found upon the wing, and strongly marks the whole tail. The comb is rather large and *convoluted*; legs clean, and of a light ash-colour. Whatever may be their origin, they are rather an interesting addition to our poultry novelties.

XIII. BENGAL. (*Class* VII. *and* III.)

These fowls are stated by the principal breeder of them, (Mr. Vivian,) to have been imported from the Indian province of that name, and were first exhibited at Birmingham in 1852; since which time that gentleman has reared them extensively— the progeny produced being (it is said) so nearly alike as to entitle them to the dignity of a new breed, although it would appear that the evident marks of a cross between the Poland and Malay families have not yet been eradicated in the lapse of time. From some cause or other they have been confounded with the Russian or Siberian, which Mr. Vivian complains of, notwithstanding that he himself gave currency to the mistake, by exhibiting them, on their first introduction, as Bengal or " *Russian Fowls*." In form they approach to that of the Malay, but are somewhat broader in the body: the plumage is usually what is termed buff, though it varies, being often mottled, and sometimes (rarely) nearly white; the legs are clean and shiny, of the same straw colour as those of the Malay. The throats of these fowls are decorated (or *disfigured*, at the pleasure of our readers) with a sort of feathery tuft.

XIV. PHEASANT MALAY. (*Class* VIII. *and* III.)

It was long supposed that this was a hybrid between the pheasant and some race of domestic fowl; but this idea having been exploded, it is now generally believed to be the produce of a cross between the Malay and the Hamburgh fowls. From the erroneous supposition above referred to, they were denominated the " Pheasant-breed; " whilst their present name originated from the pheasant-like markings of the plumage, or (which is more likely) it was given them after the Golden Pheasant or Hamburgh Fowl (undoubtedly *one* of their progenitors) already described *ante*, page 159, and which would account for the designation of " Red-Moon Pheasants " still applied to the Pheasant Malay in some parts of this country. Like the real Malay, the Pheasant Malay is long in the body, (though somewhat fuller,) has a long neck, with serpent-like head, and stands high on the legs, which also are clean, smooth, and of a creamy white shade; the comb is small in proportion to the size of the bird, but the tail is not full, but long, flowing, and sickled. The colour of the plumage is dark red, and black on the body; neck hackles long and of a lustrous blue-black shade, with (sometimes) delicate streakings of reddish-brown, — the breast being in such cases similarly marked; tail black. The hen is a much prettier bird, exhibiting in a stronger degree the pheasant-like colours and markings or spanglings in the plumage : tail black or dark brown, carried erect and flat, but the two longest feathers are slightly curved; comb very small; blue-white car-lobe; legs same colour as those of the cock, but not so long.

XV. DWARF OR CREEPER. (*Class* X.)

This fowl is the *Coq Nain* of Buffon, and *L'Acabo* or the *Coq de Madagascar* of Sonini. It has been supposed that Pliny alludes to it under the title of " the Hadrian Fowl;" and if indeed such were the case, it would go far to establish the existence, or at least the knowledge, of a pigmy race long previous to its supposed introduction in the sixteenth century, from the city of Bantam.

The Dwarf or Creeper would seem to be a stunted specimen of the Bantam breed, scarcely larger than a pigeon of tolerable size: the legs (which are feathered) are so remarkably short that the wings and rump almost trail upon the ground,— and hence their name of *Creeper*. This is perhaps the fowl alluded to by Sir W. Jardine, in his *Naturalist's Library*, where, in speaking of the Bantam family, he mentions a " still more dwarfed race, (*Gallus pumilo*,) which is extremely diminutive." They are very rarely met with now, and are held no longer in any esteem. Many varieties are spoken of by those who have visited the Philippine Islands, where they had been introduced from Camboja, a Malayo-Chinese territory, situated between Siam and Cochin China.

XVI. JUMPER. (*Class* X.)

Buffon mentions a breed of fowls in Brittany, apparently very similar to the preceding, with such abortions of limbs, as to be almost useless for walking, the movements of the fowl more resembling a sort of spasmodic jerk or jump, which has obtained for them the name of *Jumpers*. We have no knowledge of the breed, which is very probably nothing more than a misformed stunted Bantam. Aldrovandi describes a " pigmy hen," called a " Kriel " by the Germans, which " exist here and there," and " creep along the ground by *limping* rather than walking," and which evidently resembled the Jumper in the peculiar and awkward action or gait.

XVII. GROUSE-FOOTED DUTCH. (*Class* X.)

There is a breed of small fowls, which has been brought from Holland, which, though not much larger than a good-size Bantam, has a very broad, plump body, fixed upon remarkably short and feathered legs; they do not however exhibit the deformity in those limbs which characterizes the Creepers, and, unlike them, they are rather shy, and lively and active in their movements. The plumage is dark brown streaked or marked with ochre, and the general appearance so much resembles the grouse, that some persons imagine that there exists some affinity between the tribes: we regard them as from the same stock we formerly derived the old feathered-leg partridge-coloured Bantam.

XVIII. AFRICAN BANTAM.

The Americans have a diminutive fowl, which in size " will compare with a pigeon," but " symmetrically formed ;" the plumage is black except in the breast, which is spotted or mottled with a golden-yellow colour. They are perhaps produced between a black Bantam and a Nankin one—or a spangled Hamburgh.

XIX. BUENOS AYRES BANTAM.

There is a South American breed, said to be from Buenos Ayres or Brazil: they are small, and roost on trees; the feathering resembles that of the partridge, being similarly marked and streaked.

PECULIAR AND NEW BREEDS.

I. RUMPKIN OR RUMPLESS FOWL.

THERE can, we think, be little question about the *distinctness* of this very singular variety. It was known and described two centuries ago by Aldrovandi, under the name of the " Persian Fowl." They are said to exist wild, and it was long supposed that they were aboriginals of the island of Ceylon, but it appears that they have been introduced there from Cochin China,—and certainly there is something more than a family likeness between them and the Cochins. They are said to be found in Poland and Northern Germany, and are very commonly reared in Scotland. The extraordinary peculiarity by which they are distinguished is the total absence of any tail, or (as some will have) of any rump whatever, —an appearance caused by the fact, that in this fowl the last or terminal vertebra is missing, from which, in other species, spring the caudal or tail feathers. The Rumpkin is of tolerable size and of very variously-coloured plumage—some being white, grey, or black, and others of a golden yellow and speckled brown and white. The back seems to be arched or rounded, somewhat like the Guinea Fowl's, sloping rapidly from the centre of the back to a sort of round or oval extremity, where the customary *finale* of fowls are. In the wild specimens the comb is said to be like that of the Malay, but in those which are common here it is usually double, but not large; and instances occur in which a single indented comb makes its appearance. Altogether the Rumpkin would be a very handsome fellow if—*he had but a tail!*

II. FRIZZLED FOWL.

This fowl is not unfrequently called " Friesland," under the popular but erroneous idea that it has been brought from that country : it is, however, thought to be a native of Batavia in the island of Java ; and is found in Japan, Sumatra, various parts of Southern Asia, and in the Philippines. It derives its name from the singular characteristic plumage, which consists in feathers curled or " frizzled " upwards, and pointing the reverse way of those of ordinary fowls : this description applies to every part of the body, with the exception of the wings and tail, the feathers of which seldom present anything beyond a very ruffled appearance. The prevailing colour is white, though it is stated that there are specimens of almost every variety of shade, from a grey or dun to a brown, and even black. The breed is scarcely so large as some of the Game variety. The feathers at the back of the head form a sort of erect crest over an ambiguous comb, indented, and of dual form, connected at the base, but extending laterally apart. The tail is full, with two or three scimetar feathers surmounting the flat ones; legs unfeathered, and of a bluish or leaden shade. They have also been exhibited as Italian, Indian, and Egyptian Fowls.

III. SILKY CHINESE FOWL.

Although the singular breed of fowls, whose peculiar " silken " plumage so widely distinguishes them from other varieties, has long been known, the greatest confusion exists as to the division proper to be made between them. As far as our own observation extends, there are two perfectly distinct races having the remarkable feathering alluded to,—the one, in other respects, resembling the common breeds; and the second differing from the first somewhat in the plumage, and in addition, also, presenting the remarkable anomaly of having perfectly blackened comb, wattles, skin, and even, it is asserted, flesh and bones: and in this view we propose to notice them separately.

The first division of " Silky " fowls are evidently of a sub-family, not very dissimilar from the Silky Cochin or Emu Fowl. They are stated to be importations from China and Malacca, which countries are their original habitats. They are by no means common even there, and the natives are said to keep them in cages as curiosities, to sell to Europeans. The plumage is of a singularly curious texture,—the webs of each feather being quite detached and separated from each other, just as so many hair-like and distinct plumes, or long coarse silken filaments, clothing the whole body, as it were, in a dress of downy hair rather than feathers. This fowl was certainly known to Aldrovandi and other old writers, and was no doubt the original of the fabulous birds " that instead of feathers *have hair like cats*," spoken of in their works. A striking instance of the ignorant credulity which gave rise to the superstitious belief in the reality of the existence of such wonders, was afforded at Brussels, in 1776, when a bird of this identical variety was publicly exhibited to admiring and open-mouthed crowds of spectators as " a Rabbit-Fowl," alleged to have been produced from the first-named animal and a common hen! It is asserted that the naturalist Buffon gravely entered upon the task of demolishing this fable by scientific argument. The Silky China Fowl has beautiful white plumage generally, with white skin beneath, like the common breeds; comb and wattle of moderate size and deep crimson colour, the former most usually (though not invariably) single and indented on the upper ridge, with frequently a small tuft of feathers projecting from the back of the head; the hackle feathers rather fine and lengthy; tail feathers full and sickled, but by no means long; and a peculiarity we have observed in the *white-skinned* Silk Fowls is, that the bill and shanks of the legs are of a light yellow, or creamy shade of colour.

The sub-variety, recently shown as the *Algerine* seems to be in no way different from what was formerly known as the *Yellow* or *Nankin Silky* Fowl, with the substitution of the new and inappropriate name. They answer to every particular of the preceding description of the white variety, —except as regards the *colour* of the plumage, which in the Algerine is of a dusky mixture of yellow, brown, and black—very nearly approaching that of the so-called Emu Fowl.

IV. SILKY JAPANESE, OR NEGRO FOWL.

This fowl, which is an importation from Japan, is perhaps the most remarkable acquisition we have made. It is distinguished not only by its peculiar silken plumage, but also, as previously observed, by a perfectly blue-black or dark livid hue, which pervades the comb, face, wattle, skin, and bones ; some have asserted that even the *flesh* is quite black, whilst others, on the contrary, say that the flesh is as white as that of other fowls,— for our own part, we admit our ignorance of this fact, however it may stand ; and perhaps much confusion might have been avoided on this and many other points, if others had been as candid. The form of the Negro Fowl is small, compact, and square,—altogether very Cochin-like, the similarity being increased by the almost tail-less extremity of the fowl. The plumage is of snowy whiteness, or of deep but rusty black,—the former colour being by far the more commonly met with ; the feathers, though disunited in the webs exactly as the Silky Chinese, have certainly a different appearance, in consequence, perhaps, of the webs being finer and more downy ; thus instead of their looking like hairy filaments, the plumage rather compares to lamb's fleece or wadding ; and it is curious that, although at the present day the two sub-varieties are confounded together, the old poultry writers clearly perceived the difference in plumage,—for whilst, as we have seen, they speak of the Silky as having " hair like cats," they likewise notice " wool-bearing fowls," that are " covered not with feathers, but *with wool like sheep*." When pure bred, the comb of the Negro Fowl is of a dark purple colour, double, and in the form of the Malay: it is surmounted by a small and scanty tuft of feathers; large dark blue or purple ear-lobe and cheek-patch, with small blackened wattles; hackle feathers thick and downy, but much shorter than in the Chinese variety; the quills of the wings, according to Mr. Baily, are naked like those of the porcupine; the tail is perfectly black, and the legs also,—the latter (which are very short) being covered by a remarkably hard, smooth, scaly substance, contrasting strongly, in the *white* Negro Fowl, with the thick downy-clothed thighs as round as a ball; the shanks are heavily " booted " on the outer side with feathers of the body colour.   The hen, though small, is not much less in size than the cock: the comb is not so much developed, but the crest more so than in the male.

The *Black Negro* bird, as may be imagined, is altogether black from tip to tail; but in every other respect identical with the white.   Very few specimens of this kind have, however, found their way to this country as yet.

NEW BREEDS.

V. PTARMIGAN FOWL.

LITTLE is known of the history or habits of this exceedingly novel and handsome variety: even

o

the whereabouts whence they have been imported cannot be indicated, beyond a "belief" that they were brought some four years ago from the "north of Europe,"—certainly indefinite enough; but perhaps they may be natives of Siberia. They made their "first appearance in public," at the Metropolitan Show in July, 1853, from the yard of their original proprietor, Dr. Burney, of Brockhurst Lodge, near Gosport. From some little resemblance which they bear to the Ptarmigan fowls in their winter dress when roaming over the Cheviot range, this new breed has been named after them, which, from the confusion it will no doubt hereafter create, is not satisfactory: and still less so is that of "Grouse-footed Poland," with which variety there is not the slightest likeness beyond the fact of the so-called Ptarmigan fowls having a tuft of feathers on the head, and even that is of a totally different character to the Polands. They seem very timid and shy in their movements; and from this and their anti-granivorous habits, (preferring insects and ants' eggs,) we are inclined to suspect some connexion with the grouse tribe. The form of this fowl is elongated, though the proportions are certainly diminutive: back very flat, and the neck straight and erect; the tail long and flat, carried also elevated in a line with the head, which latter is surmounted by a tuft of feathers, something higher in front than behind—not much unlike the peculiar shape of the caps worn by the peasants of Normandy and Picardy; it has also a small comb, like two single combs joined together at the base. The legs are, in proportion to the bird, very long, but it does not look so awkward in consequence, as most long-legged fowls do, from the thighs being thickly feathered quite over the hock joint, after the fashion of the falcon and vulture tribe, the shanks of the legs also being feathered down to the very toes. There is only a slightly pinky face, and incipient wattles. The hen is very little smaller than the male; but the tufted crest being better furnished, assumes a round, ball-like form, and, added to a much more pointed tail, gives the hen a rather different appearance to the cock. The plumage of both consists of pure snowy-white feathers of singular beauty.

VI. PRAIRIE FOWL.

This singular breed was imported during the spring of 1853, by the Messrs. Baker, who obtained them from an American gentleman, as their name indicates. These fowls inhabit the immense prairies of the New World, where they are found in great numbers, affording excellent sport, though they are exceedingly difficult to capture alive. They are of rather diminutive proportions, but plump and rounded forms: neither male nor female have any description of crest or wattle; the

former has a broad, expanded, fan-like tail, resembling in shape (though of course on a much
reduced scale) the tail of the turkey. Although
this variety of fowls seems to belong, like the preceding one, to the genus *Tetraonidæ,* or grouse
tribe, and as such have nothing to do with " Domestic Poultry," our readers will perhaps pardon
the introduction into our pages of the following
brief description of them, (the only one which has
yet been published,) which appeared, together with
an illustration, in the *Field* newspaper of June
25th, 1853.

" They resemble the grouse in colour, and in
being feathered down the legs, but are quite
double the size; the male bird has eight or nine
feathers, growing down from each side of the
head, and hanging forward between the shoulder
and breast. But in the breeding season, and
when in attendance on the hen, he frequently
raises and brings them forward in a line with the
eye, gracefully bends his head, swells out his neck,
and makes a singular humming or booming noise
—mostly towards evening—which may be heard
at a considerable distance.

" The hen differs only in having the neck feathers much shorter, and being more faintly marked
[and the tail flatter and squarer]. They appear
very hardy, and we have no doubt they would
breed and thrive well here—as they have had the
same food as pheasants, and are much improved in

condition since their arrival." A pair of the above fowls has been recently added to the Marquis of Allen's collection.

VII. CEYLON JUNGLE FOWL.

Of this variety, at present but little known, some birds were imported from the island of Ceylon, during the year 1852, by a gentleman at Bristol, from whose possession they have now passed into that of Mr. Brissell of Birmingham. They present many points of similarity, both in form and character, with the Cochin Chinas; than which, however, they are very much smaller, being scarcely so large as some specimens of the Game breed. From the description which has been published by the present proprietor, we gather that there are two (probably more) sub-varieties—*silver* and *golden*. The silver has plumage of a clear white ground colour, *double* shaded, or laced with black, differing from the ordinary lacing or spangling in having *two* rows of elongated oval markings, about an eighth of an inch in breadth each, running round every feather, which has thus, first an outer margin of white, then a narrow line of black, succeeded by a similar one of white, and equidistant therefrom a second black lacing, the shaft of the feather being, finally, white,—these markings (as is always the case in fowls of the spangled varieties) being more beautiful and regular in the female.

The golden race have precisely similar lacings, but upon feathers of a light ochrey-brown shade for the ground colour. The characteristics of the variety are, a small head, single comb, loose and puffy red cheek-patch, stunted and flat tail, with very short and straw-coloured legs. Supposing this breed to be perfectly distinct from any of our own continent, it is still doubtful whether they would ever become sufficiently acclimatized here to perpetuate their race.

VIII. EGYPTIAN FOWL.

At the Hitchin Poultry Show, held in 1853, there was a pen of white-plumaged fowls entered as "a new and distinct breed imported from Egypt." Of this we have no knowledge beyond the fact that it is "a compact short-legged fowl, with a peculiarity about the legs and back," transmitted in the chickens, "that seems to indicate a distinct kind."

IX. BUFF POLANDS.

In concluding this portion of our history of the gallinaceous family, we may mention that since our notice of the Poland variety, some fowls have made their appearance at the different exhibitions, which *are said* to constitute a new sub-variety,— the plumage is of a yellowish or buff shade entirely, including the crest, tail, and beard, with a slight mottling of white on the edges of the feathers, where in the golden-spangled Polands the black spangles are seen; the flat tail-feathers are likewise indistinctly edged with white. They are produced from the yard of Mr. Vivian, who has given his attention more particularly to the Poland family. We have no desire to detract from the merit or rarity of the novelty we are speaking of, but cannot avoid thinking that our remarks upon the probable origin of the fowls termed *yellow-spangled*, may also perhaps account for the production of the Yellow or Buff Polands of one shade.

X. RANGOON.

This fowl, of which we have not yet seen either a specimen or description, has been exhibited at the Birmingham Poultry Show, held December, 1853; and although we thus include it under the list of " New Breeds," we would observe, that it may possibly be only a variation of some already existing and known variety.

XI. HORNED ATLANTIC FOWL.

Another (we presume) recent introduction to this country; of which, having no knowledge beyond the fact that it was exhibited at the Southampton Show, November, 1853, we can only say, that our remarks as to the *identity* of the preceding novelty must apply equally to this.

CHAPTER VI.

DESCRIPTION OF THE PHEASANT—PEACOCK—TURKEY—
GUINEA-FOWL—PIGEON—SWAN—GOOSE—DUCK.

WE now enter upon the history of an assemblage of fowls, comprising some which may, at the first glance, be thought scarcely to fall within the scope or design of our work,—such, for instance, as the pheasant tribe, kept, or (to use the orthodox phrase) " preserved," principally for the amusement they afford to the sportsman—or, like the gorgeously-clad peacock, now regarded only as beautiful objects to please the eye of the spectator, and to ornament the seats and pleasure-grounds of the nobility,—or such, again, as the pigeon family, perhaps, more correctly speaking, birds of the aviary rather than inhabitants of the poultry yard ; but to those who more attentively consider the subject, it will, we think, be apparent that they all are more or less connected with that branch of rural economy under which are included our Domestic Fowls, and of which we have undertaken to treat.

PHEASANTS.

Sub-genus *Phasianus*.

I. COMMON AND RING-NECKED PHEASANT.

Origin and History.—The Pheasant is generally believed to have been originally, and at a very remote period, brought from the countries bordering upon the Phasis, (whence their name was probably derived,) a river in Asia Minor. The Argonauts, in their celebrated expedition to Colchis, together with the golden fleece, conveyed back with them the Asiatic Pheasant, the plumage of which was equally rich and resplendent with the fleece. This bird, indeed, may successfully vie with any other of the feathered tribe, for beautiful symmetry of form and shape, as well as for the rich and luxuriant plumage, or the varied and dazzling gaudiness of its colours. It was distributed over every part of the Old Continent—even to China and the borders of the Tartar empire. In the present day it is met with naturalized in almost every clime—in Siberia, at St. Helena, in North America, and in our own country, where they are abundant, as likewise in France; in the latter country they are more carefully and extensively reared than with us. Pheasants are mentioned in England as early as the 13th century.

Description.—Formerly naturalists regarded the

common Pheasant of Britain as a sub-variety of the beautiful Ring-necked Pheasant of China; but by some they are now looked upon as two different races, or varieties of their tribe,—though their *distinctiveness* is by no means positively determined. The general characteristics of the Pheasant tribe are thus described by Jardine in his *Nat. Lib.* vol. xiv. The bill strong, curved on the upper ridge, and naked at the base; head clothed with feathers, except some considerable space around the eyes, which is covered with a naked* varicose or granulated skinny patch, usually of a red colour. Wings short and rounded; tail wedge-shaped, and of extraordinary length, frequently, in the males; the anterior toes in the feet united by a membrane to the first joint; the males possessing short strong spurs. These points are all present in the Common Pheasant, the plumage of which is exceedingly beautiful, of a glossy light reddish-brown, resembling copper bronze, and variegated with regular markings, or spanglings of black over the whole body: the breast exhibiting reflections of metallic purple hues—also seen sometimes upon the wings; the head and neck

---

* Temminck asserts this to be an error; and states that this skin is of a thick expansive texture, and covered with a number of "loose filaments, or minute bristles, forming a close velvetty tissue." He further asserts that the pheasant has the power of distending his skin, and imparting to it various shades of red,— more especially in the season of love, or when, as in the case of the turkey, it is excited or agitated.

displaying feathers of a brilliant bluish-green, with metallic reflections. In length it extends, from tip to tip, to three feet. The female is of smaller size, and has (comparatively speaking) a short tail; the feathering is of a sober brown or dark chestnut colour; the black markings upon the body and tail being much more distinct than in the males, except indeed upon the breast, which she has perfectly clear from spangling: a small band or tuft of short, thick, black feathers just below the eyes, also distinguishes the female bird.

The *Ring-necked* does not differ materially from the preceding, but is of rather a smaller size, the extreme length never exceeding (and but seldom reaching) two feet and a half. The black barring on the plumage is somewhat different and more delicate; but the principal characteristic mark, however, is the white ring which encircles the bottom of the neck, narrowest at the back and front, but expanding into a broader band in the sides of the neck. Both this and the preceding principally inhabit the forests of China, but are said never to intermingle.

II. BOHEMIAN PHEASANT.

History.— This breed is very scarce in this country, specimens being rarely met with: it seems equally unknown on the continent, having escaped the notice of the French naturalists. It is very doubtful whether the country whence the bird takes its name, is its original habitat; and the entire absence of any information as to its origin or history has induced many to regard it only as an accidental variation in the Common Pheasant, in which the colours are run out or weakened by defective secretions. (*Yarrell.*) They have, it is stated by Jardine, of late years become " extremely common " in Scotland, though upon what foundation the opinion has been given we know not.

Description.—In size, form, and plumage the Bohemian closely assimilates to the Common Pheasant, but upon the neck, breast, back, and wings, the feathering is of a light buff shade, or as Mr. Baily expressively describes it, the plumage looks as if " *it had been washed over with cream.*" From this buff or creamy shade, the black markings assume a more distinct appearance than in the Common Pheasant, arising, perhaps, from the greater contrast between the colours. Owing to this singularity of plumage some have conjectured that the origin of the bird may be traced to the union of a Silver with a Common bird,—but for various reasons this does not seem probable. The *female* Bohemian is described as of much smaller proportions than the female of the Common Pheasant.

III. SILVER PHEASANT.

History.—This variety, which is said to be the most powerful of its tribe, is a native of the northern provinces of China, whence it has been imported into various parts of Europe. It was first noticed in this country by Albin, about the year 1737, when it was figured in his work under the name of the " Black and White China Pheasant,"—by which it is sometimes designated by recent writers.

Description.—In size the Silver is generally described as being less than the common variety,—although, from our own observation and inquiries, we should say that it is certainly quite as long and large: the plumage when young is of a russet grey or fawn colour, clouded or indistinctly shaded with brown, the under-parts being a creamy white; but in the adult bird the colours become clear white and black. The head is surmounted with a loose tuft of glossy purple-black feathers, and the throat, breast, and under-parts are of a rich metallic blue-black; the back of the neck clear white; so also the back and wings, though likewise pencilled or shaded diagonally with black. The tail is long and convex, or rounded at the commencement; it consists of long plumes arched upwards, —the two upper feathers being marked in a similar manner to the back, and the two middle ones are long and flat, and entirely white. The eye is surrounded by a skinny substance, of large size, and of similar texture to that described on the Common Pheasant; but in this variety it is more developed, extending over the eyes almost like an incipient comb, and drooping below as a small wattle. The legs are of bright lake shade, and furnished with sharp white spurs. The female is much smaller than the male; the feathered crest on the head scanty, and of a deep brown hue; the eyes encircled with the peculiar tissue of velvet-like filaments, but in a much less degree; tail shorter, considerably less elevated, and straighter, the central feathers brownish, and the lateral black and white; throat feathers of a creamy shade, and the neck and under-parts dusky white and brown with irregular bars of black; the lower part of the back of the neck, with the back itself, and the wings, a grey, streaked with brown and (more delicately) black.

IV. GOLD PHEASANT.

History.—Like the preceding, this variety inhabits the forests of China, where they are in great number. They are there called " Gold-Flower Fowl." In Europe they are tolerably well known, especially in Germany, where they are rather extensively reared. In England, it would appear, from Albin's work, that they had been imported some time previous to the year 1735, as that writer then depicted some birds evidently of that kind, under the name of the " Red Pheasant from China," and before this, if known, they were passed over by Aldrovandi and other old authors. Cuvier suggests that this, the most gaily and splendidly attired of its kind, was the origin of the fabulous creation of the Phœnix among the ancients.

Description.—In size this is fully one-third less than the Silver; the length never extending beyond 2 feet 8 inches, of which the magnificent tail takes off nearly 2 feet. Upon the head is a rather full and high crest of loose plume-like feathers of a bright yellow shade, terminating in a point and drooping back, somewhat after the fashion of the cap worn by the ancient jester or Merry Andrew ; the eyes encircled by the skinny wattles peculiar to the pheasant, but in this variety of a livid hue; throat and neck clothed entirely round in a cape or lappet of expanded and elongated feathers, resembling the hackle feathers of the common cock, but truncated or cut off square at the ends ; the feathers are of a brilliant orange-red colour, striped and variegated with black, very much like the double flower of some marigold plants ; lower front part of the neck, with the under and hinder parts, of a rich, resplendent scarlet or crimson, deepening in shade as it descends ; the bottom part of the back of the neck a bright or emerald green delicately tipped with black: back and saddle a rich deep yellow; from the latter the feathers droop on either side of the tail, resembling hackles, but stiff and almost horny in substance; upon the shoulders steel-blue and metallic shades; greater and lesser coverts of wings of a deep brown spotted and marked with darker; the tail feathers are broad, folded together and slightly bowed or arched in the middle,—the colour on the upper portion being a very dark brown, striated and rather minutely marked with black, and the under short feathers of a " cinnamon " colour, as it is sometimes termed; the legs long, slender, and a dull yellow shade. The female is smaller, has more elongated feathers upon the head, which she has the power of raising something like a crest, the plumage throughout of a rich deep brown shade (the throat and underparts rather lighter) barred with regular black bands; tail short (comparatively) and presenting similar markings on the same ground as the body colour. When young the Gold Pheasants possess none of the beautiful plumage of the adult birds, and indeed are scarcely recognisable, the plumage being of a streaky yellowish fawn colour.

In addition to those above described the following varieties of the Pheasant tribe are mentioned by various writers:

DIARD'S PHEASANT (*Phas. Versicolor*).   A

native of Japan, and first noticed by Mr. Diard, after whom it is named: plumage of a bright golden green: size larger by 6 or 8 inches than the common variety.

BARRED-TAILED PHEASANT. A native of the sunny mountains of Surinagur, and of the confines of the Chinese empire; about the size of the Silver Pheasant, but with a tail said to vary 2 feet in length up to 7 feet! The plumage a golden yellow and black.

SŒMMERING's PHEASANT. A native of Japan; nearly similar in size to Diard's variety: the plumage of a rich copper bronze or reddish brown, marked with black.

LADY AMHERST's PHEASANT. Only known from two specimens presented by the king of Ava to Lady Amherst: it is a crested variety, rather larger than the Common Pheasant, and of beautifully variegated colours.

Some writers speak of the " Peacock Pheasant from China," the " Black and White," and " Pied " Pheasants, but of these no account is given, and it is probable that they belong to one or other of the varieties already enumerated. In the month of December, 1853, a very fine male Pheasant, of perfectly white plumage, was shot in the neighbourhood of Botham: the beautiful bird, with three others of similar plumage, had long been accustomed to visit the residences of J. H.

Amory, Esq. and other gentlemen, at Hayne, and, it is said, had become almost domesticated.

PEACOCKS.

Sub-genus Pavo.

I. COMMON OR INDIAN PEACOCK.

History.—This most gorgeous and beautiful of all the feathered race originally inhabited Central India, whence in remote ages it was introduced into Greece, and in the lapse of time its migration westward has been completed. In the dense forests and thick jungles of its native country it is still found in the greatest abundance, likewise upon the banks of large rivers, as the Ganges, Jumna, &c. ; there too, in the Ghauts, the Dukhun, and the passes of Jungletery, travellers have encountered the wild Peafowl in vast flocks numbering several hundreds. Varieties have also been met with, not differing very widely from the common or domesticated, in Burmah, Ceylon, Java, and Sumatra. After its introduction into Greece, among the ancients, the island of Samos (sacred to Juno) was celebrated for the rearing of Peafowls.

Description.—The body of the Peacock is elongated in form, and the back arched over like that of the Guinea-fowl : the neck is long but not slender, though gracefully bending like the Swan's, by that means giving a great prominence to the breast, as well as a dignified carriage ; head surmounted with a crest or " aigrette " (as it is generally called) of feathers, said to be 24 in number, and consisting of long slender shafts, quite unwebbed, except just at the tips, which terminate in a fringe of silken webs of a golden green ; the head and face covered with feathers of a bluishgreen, except a patch of loose skinny texture just above and below the eye, and of a livid bluishwhite hue : neck a dark green, terminating in a deep steel-blue or purple upon the breast, whence the plumage assumes a rather darker shade upon the under and hinder parts : golden green, blue, and dark brown blended together upon the back : wings a sort of fawn colour, clouded like tortoiseshell with dark brown and a rich maroon or mulberry shade, the principal quills and flight feathers being brown shaded with black and green. Descending from the back, and glittering with the most resplendent hues, the magnificent long plume feathers of the *train* (popularly but improperly termed the *tail*) next present themselves. These plumes, which commence upon the lower part of the back, are apparently seated in a strong muscular development or expansion *over* the rump, which gives the peacock that peculiar power of expanding or spreading its gorgeous train of glittering feathers, in a way doubtless familiar to most of our readers : this train is composed of long plumes of silken web attached to long and slender shafts—not of the horny nature of the quill, but pliable and elastic; they form the coverts for the *real tail* of the peacock, which consists of several short stiff quill-feathers, (18 in number,) which, seated in the natural place—upon the rump, serve as a ledge on which the long and sweeping train rests, and is kept from entirely draggling upon the ground. These plumes, at first of a brilliant greenish yellow, gradually assume a rich hue of copper bronze, reflecting the most radiant tints of indescribable beauty, and dotted over with numerous circular or ocellated disks, each having the centre eye, a dark blue or purple, encircled by an emerald green, with an outer margin or rim of bright yellow; the shaft of each plume terminating in two streaming filaments, which project beyond the outer edge of the expanded train like a fringe. The legs, which are very long, are furnished with sharp spurs. The Peahen, though crested like the male, is destitute of the beautiful train, having nothing beyond the usual terminal appendage: the female is likewise much inferior to the cock in regard to brilliancy of plumage, which, in her, is much softened by darker and duller shades. Our portrait (*frontispiece*) is taken from a specimen formerly belonging to Miss Morse, of Norwich.

It occasionally happens that the peacocks of this breed will come out with pied or mottled white plumage; and instances have been known in which the feathers were entirely white, the ocellations being only indistinctly traceable. Birds with such variations, though highly prized on account of their rarity, seem but the productions of chance ; whilst scarcity is too dearly purchased at the sacrifice of beauty.

II. JAVANESE PEACOCK.

History.—This variety has been met with in Java, Sumatra, Burmah, and Japan. It was first described two centuries ago by Aldrovandi, and from this circumstance is called by some modern writers the " Aldrovandine Peacock ; " and specimens having been obtained from Japan, it has also been named the " Japan Peacock; " from which much confusion has arisen, it being confounded with the " *Japanned* " Peacock—another variety until recently unknown, and with which we are still only imperfectly acquainted.

Description.—Of size and form identical with that of the common Indian Peacock, it is nevertheless strikingly distinguished from it, not only by considerable variations in the colours of the plumage, but a remarkable crest of a totally different character ; for whilst in the common variety it consists, as already stated, of naked shafts ending in an ocellated web at the tip, in the present it is much longer, and is composed of shafts webbed from the origin on the crown of the head quite to the tips, such filaments or webs extending a uniform breadth from the shaft: the patch surrounding the eye and upon the cheek, instead of being white, is a bright greenish yellow or lemon colour: head and neck a bright golden green, something like the lighter plumage of the green parrot, but tinged with coppery hues: breast and under parts of a dull dingy brown, with a greenish bronze hue pervading the whole: back spotted, and the colour a dull sage or withered leaf-green: coverts and wings dark blue softened with a tint of green, uniform throughout, except the quills and flights, which are of the same shade as the under-parts; plumes of the train less ample and luxuriant, inferior in brilliancy, having but little of the radiant colours of the common, with ocellated spots, fewer, and not so clearly defined: the legs shorter and not so slender. (*Frontispiece.*) The female has a very dull, sombre dress, but not much unlike the common peahen; the crest, however, differs as much as in the males of the two varieties.

III. JAPANNED PEACOCK.

History.—Of the source whence this new variety has been acquired we are ignorant, as no account has yet been given of its original habitat, though it seems highly probable that it is an importation from the great Eastern Archipelago, and may possibly be found distributed over the same spots as the preceding variety. Sir R. Heron, who first noticed the " Japanned " breed, in the *Proceedings of the Zoological Society*, 1835, conjectures that the breed " originated in England," grounding his opinion solely upon the fact, that some years previously it had " suddenly appeared " among the common variety in the collections of two or three noblemen with whom he was acquainted,—obviously a conclusion very prematurely arrived at. A pair of this variety were, a year or two since, (and possibly may be still,) in the Zoological Gardens, but we have been unable to learn from what quarter they were procured.

Description.—The only description of this variety that has yet appeared, is from the pen of the Rev. E. S. Dixon, who states that it is of smaller proportions than the common Indian Peafowl: the cheek-patch is white, but much smaller : the wings are blue-black edged with metallic green, and the shoulder feathers of the wings are bright jetty black, having a "*japanned*" appearance— whence no doubt its name has been derived. In this variety also, " the imbricated feathers on the back are smaller and less conspicuous, and the whole colouring of the bird is of a darker tone : the hen, on the contrary, is very much lighter than the common sort, with a tendency to spangled (perhaps even ocellated) plumage all over her body ; her size is also inferior, and her proportions more slender." (*Domestic Poultry.*)

TURKEYS.

Sub-genus Meleagris.

Name and Origin.—The Turkey, it is said, was seen in America by the first discoverers, and was named by the Spanish doctor, Fernandez, *Gallus Pavo* and *Gallus Indicus*,—the Peacock of the Indies. It was found both in a wild and domesticated state on the arrival of the adventurers. There seems little doubt that it was first procured by the Spaniards, shortly after they made the discovery of the New World, from some of their possessions then acquired there; though it was, perhaps, previously domesticated among the Mexicans. In 1525, Oniedo, a Jesuit traveller, (who resided for many years in the West India Isles,) published a *Nat. Hist.* of the " Indies," (the name given to that portion of the American continent situated between the tropics,) in which he describes the Turkey as somewhat like unto a peacock; adding that it was met with in vast numbers in New Spain, but had been conveyed thence to the Islands and the Main, and subjugated by the colonists from the Old World. Although the Turkey was not received in France until some years subsequently to the above date, it is tolerably clear, from the name given to it, *Dinde* (d'Inde), that our continental neighbours obtained it from the same quarter,—namely, the West Indies. The indirect source whence we have derived our stock is, however, by no means so apparent; the appellation by which it is known amongst us, serving only to obscure and mystify it: as with the Poland fowl, so with the Turkey,—the name has proved a fruitful field for conjecture, and our poultry writers have given full scope to their "fancy free," in speculating upon the origin of the English synonym. The best explanation that can be offered, we believe, is this: that as the Turkey is known to have been first received from Spain (after its introduction there) by the Italians and other nations of the South, it is highly probable that the fowl was imported hither by some of the Levant traders, when its identity became lost under the general designation of "Turkey,"—a prefix at that period indiscriminately applied to coffee and various other (then) novel importations, although not the produce of the country so named, nor indeed in any way connected with the East. The generic term (*Meleagris*) by which modern naturalists distinguish the Turkey, though of course it cannot be said to be absolutely *improper*, does not seem very appropriate, and formerly caused the Turkey to be confounded with the Guinea-fowl,—*Meleagris* being the name given to the latter bird among the ancients, to whom the Turkey was quite unknown.

WILD RACE. (*Meleagris Gallopavo.*)

History.—The wild Turkey is scattered over the American continent, from the extreme north-western portion as far as the Isthmus of Panama, south of which M. Buonaparte assures us, in spite of what other naturalists assert, the bird is never seen. At the present day they are principally met with in the dense woodland tracts upon the Mississippi and Missouri, and in the uninhabited districts of the Ohio, Kentucky, the Illinois, &c. In Virginia, once celebrated for the fine breed which abounded there, they are becoming scarce. In Canada also the Turkey was formerly very plentiful: but, as in other parts of the New World, since the planting and cultivating of extensive regions and tracts, consequent upon the wood-clearings of the settlers, the wild breed has been gradually driven, by the progress of population and civilization, into undisturbed and unexplored districts, and are therefore now becoming rare; migrating in excursive flocks into the native forests. In the Indies they have been extensively reared by the Portuguese: the natives make elegant articles of clothing and fans out of their brilliant feathers. In the state of Louisiana, it is said, the French settlers manufacture umbrellas from the plumes.

Description.—The Turkey of nature is described

of much more magnificent appearance than the domesticated descendant. It is also a much larger bird, though not so " lumpy," but more symmetrical in form than the latter: the weight may perhaps be stated within reasonable limits at 35 lbs. or 40 lbs., and though no doubt specimens may *occasionally* have been seen exceeding this standard, they are only extraordinary cases, and we should hesitate merely on that account to place too implicit faith in the " traveller's tales " that are told of fowls of 60 lbs. weight. The plumage of the wild Turkey, though of various shades and brilliancy, from a metallic black and brown to a bronzed hue and almost rainbow tint, is always superior to that of the domestic. In the coloured sub-variety the back and breast is of a light but rich reddish brown, with darker streakings: the wings, which are longer and more depressed or pointed downwards than in our own kinds, are of a bright greenish or parrot-like yellow, beneath two rows of feathers of a duller green, but bronzed with a red or copper hue, and the lower portion white, streaked or pencilled with black: the tail displays various colours blended in successive rows of feathers,—brown, black, yellow, sea-green, slightly flaked with white, and presenting the radiant pinky and bluish tints seen on some marine shells: the legs longer and stouter than the common, of a pinky or flesh colour, with short spurs: the head and upper portion of the neck covered with a loose naked skin of bluish livid hue, continuing downwards, for two or three inches from the base of the upper bill, into a pendulous fleshy tube, resembling an elephant's trunk in miniature, and of a mottled red and white colour: under the lower bill and throat the loose skin from the head collects together in loose scarlet folds, and hangs like a small wattle: from the breast depends a tuft of dark feathers, of hair-like texture: legs rather long and stout, of a fleshy colour, and furnished with strong thick spurs. (*See plate* V. *fig.* 3.) The female is said to be of much less brilliant colours, the predominating shades being brown fringed with black: she is also much smaller, and has very little of the naked skin upon the neck or head, the feathers extending higher up than in the male.

HONDURAS TURKEY. (*M. Ocellata.*)

The existence of this beautiful variety of the wild Turkey, is only known from a single specimen captured some years since by the crew of a British vessel, at that time taking in timber from Honduras Bay. It was forwarded to Sir Henry Halford; but meeting with an accident in its transit up the London river, it died, when it was presented to Mr. Bullock's Museum, in Egyptian Hall; and upon the dispersion of that splendid collection, the beautiful specimen found a resting-place in the

Paris Museum of Natural History. From the description given of it by M. Temminck, it must be a superb fowl in its living attire. It is somewhat smaller than the *M. Gallopavo*, from which it differs but little except in the great resplendence of the plumage, in which is blended the greatest variety and brilliancy of colouring, surpassed only by the Peacock: in its feathers are blended together shades of gold, bronze, red, green, blue, brown, and black, a radiant metallic hue pervading the whole: the markings, which upon the back are regular bars, assume a rounded appearance upon the tail coverts and tail, and become like the eyes or ocellations upon the Peafowls—whence it therefore derives its scientific name. We have no record of this variety being met with in any other part of the New World, though it might perhaps be found inhabiting the unexplored regions of Central America: it is, however, exceedingly rare. We have seen it stated that a fine specimen is now in the British Museum, but we have no account of any second importation to these shores.

DOMESTIC RACE. (*M. Gallopavo.*)

This descendant of the native Turkey of America closely resembles its original, although it is of rather inferior proportions,—and whilst often attaining considerable beauty of appearance, its inferiority in regard to plumage is unquestionable, the effect perhaps of degeneracy consequent upon domestication; in which state it is now found distributed over almost every portion, or at least every civilized portion, of the globe; and it is a remarkable proof of the degeneration which this fowl has undergone, that it is not alone in Europe that this inferiority is seen, but even upon its native shores the contrast between the wild and domestic race is very apparent. The weight of the latter is usually as much exaggerated as that of the former has been: if we give 15 lbs. to 20 lbs. for the male, and 10 lbs. to 14 lbs. for the female, (both living specimens,) of the first year's growth, we should be fixing a full high standard average; and 36 lbs. would be an extreme weight for aged fowls: as for the intermediate weights, he would be a fortunate man who, by ordinarily good feeding, could add 5 lbs. weight to his Turkeys for every year of their age. The ample description given of the wild progenitors would seem to render it a work of supererogation to enter into any minute details respecting the domestic denizen of our yards: but the following particulars on some of the principal points may not be out of place, nor indeed altogether unacceptable, as being deduced from more extended and accessible observation. The breast should be broad, ample, and prominent; the wings large, strong, and well rounded on the outer sides; the loose membrane, or thick wrinkled skin or hood, which covers the head and neck, should be studded with red pimples or nipples in front; and when the fowl is excited by any kind of passion, (either love or hate,) the changes in the colour of this hood and frill on the neck, from pale livid blue to white and red, and thence to purple and deep crimson, should be both rapid and frequent, as denoting high and vigorous health. The trunk-like extension of the skin from the forehead increases with the age of the fowl, until, in some aged specimens, we have seen it covering the bill, and drooping some 6 or 7 inches below; independently of which the male has the power of stretching or elongating this appendage at pleasure. The scarlet wattle-like membrane which hangs immediately below the throat, should in full-grown fowls hang about one-third down the neck, and is usually, in health, flaked or streaked with white: the hard, horny hair-like feathers on the breast at the bottom of the neck seldom depend lower than 5 or 6 inches, though it is quite possible that in very old birds they may grow an inch or two lower. In the lighter coloured sub-varieties the bill, we have noticed, is usually of a dusky horn colour, whilst in the darker or black it is of a creamy and yellow shade; upon the upper portion of it is seated the aperture of the nostrils; and immediately behind the eye are seen the openings of the ears, partially covered over with small ruffled or frizzled feathers: the wings each contain twenty-eight quill feathers, and the tail eighteen,—or, more properly speaking, the *tail coverts*, for that fan-like portion which is capable of expansion, like that of the Peacock, properly form the coverts for a smaller and in-ferior tail of straight horizontal feathers, by which the former are supported.   The hen, in addition to her smaller proportions, is easily distinguished by the naked skinny portions of the head and neck being lessened by the feathering (though scanty) coming higher up,—also by the trunk-like termin-ation of the former being very much smaller and not capable of elongation, by the absence of any pendant wattle under the throat, and by the greater number (generally) of pimples upon the neck ;—and finally, it is said, by her inability to expand the tail coverts; although this last point is disputed by the Rev. E. S. Dixon, and as there seems an insuperable legal difficulty in the way of *proving a negative*, our readers must exercise their own discretion in accepting whichever *dictum* they please.

SUB-VARIETIES.

Although perhaps not so distinctly marked as in the common cocks and hens, domestication seems to have split up the Turkey tribe into nearly as many sub-varieties, distinguished by (almost ex-clusively) a difference in the colours of the plumage alone. The nomenclature of these, however, is by no means satisfactory: thus we have the "Norfolk," (so called,) made to include all black-plumaged fowls; "Cambridge," all greys and black and whites; "Shropshire," chiefly browns, &c.; all of which *colours* are as much the *exclusive* distinction of Essex or Surrey, as of the first-named or any other counties: for, by the same rule, we might run through the forty counties of England, and possess as many sub-varieties: but a Norfolk *black* Turkey is no more distinct from a black of the county of Surrey, than a herring caught at Yarmouth is distinct from one pulled up off Brighton. The prefix of the county may perhaps have been originally, and with more reason, applied to the *dead* poultry, as from the superior way of fattening or "getting them up" for market, in the district whence they were sent, that mode of expressing their *quality* was adopted: but we see no reason to make any other distinction in separating the various sub-varieties of Turkeys than that universally followed with other poultry,—namely, the difference in the colours of the plumage. We shall then have the following:

WHITE (pure).
BUFF, or Fawn.
COPPER, or Red; including the so-called "Dutch."
BROWN, or Bustard Breed.
BRONZED, or American.

BLACK (Norfolk and Cambridge).
BLACK AND WHITE (Cambridge and Duke of Rutland's).

White.—Some will not admit that this is any sub-variety at all: the original race of wild Turkeys certainly is said to have been black; and yet one would suppose that the white must also have been a primitive colour with them,—or else the transition from black to pure white would be most extraordinary and unaccountable. In America the white Turkey is exceedingly rare; at the great Poultry Exhibition there in 1849 there was only one solitary Turkey of that colour, sent from Providence. On the continent, in France, the whites are very extensively bred, and held in high esteem, on account, as the good-wives assert, of their being easier reared and sooner fattened than any other kind; in Languedoc, Provence, &c., they are very generally (almost exclusively) kept: but in the Dauphiny all kinds are equally esteemed, and Turkeys of every shade reared. In our own country they are but seldom seen, as they are considered less hardy, and in every way inferior to the darker; but the wide-spread prejudice which exists here against *all* kinds of *white*-plumaged birds of every variety, may probably have tended to discourage their being kept, without any sufficient reason or experience. All are agreed, however, that White Turkeys, with their snowy plumage, red hood and frill, and black ermine tuft on the breast, are remarkably handsome fowls; and therefore not unworthy to be reared, if not for their value as poultry, at least as ornamental fowls for the park or pleasure-ground. They are said to have been originally introduced from Holland, but whence our Dutch neighbours in the first instance procured them is not shown. Occasionally some inferior white specimens, streaked irregularly and slightly with black on the wings, back, tail, &c., are swept into a sub-variety, called "Cambridge Greys;" but these, as we have observed, appear to be only imperfect whites; whilst the more distinctly marked we describe subsequently under the Black and White sub-variety.

Buff.—These fowls, with plumage of a light buff, or delicate fawn colour, have an exceedingly neat and quaker-like appearance, when the shade is unbroken,—but it is sometimes broken with white. This colour is but rarely met with now.

Copper, or "Dutch Red,"—supposed to be, like the White, immediately procured from Holland; and formerly held in great esteem. Some of the darker shades are called "Cinnamon," being of the dull reddish hue of the sable hair, or like the colour of the hair of a tortoise-shell cat. This sub-variety is now met with in Cambridge and Essex alone; and does not seem sufficiently prized to lead to its being cultivated to any extent. At the last Eastern Counties' Show, a pen of the darkest fowls were entered under the designation (perhaps more expressive than elegant) of "liver-coloured."

Brown.—This sub-variety, better known as "the *Bustard breed*," appears to be common in

Cambridgeshire. The plumage is of a dark brownish red, the webs of the feathers being pencilled
and tipped (irregularly) with a darker olive brown
or rusty black, and mottled or broken, especially
on the back and roots of the tail coverts, with a
greenish grey and white. They are fine birds and
of large make. The term " Bustard breed," may
perhaps have been applied to these Turkeys from
the resemblance of the feathering to that of the
Bustards, which were once very common in Norfolk and Cambridgeshire; but having now disappeared, some floating popular idea may even exist
that the Bustard has really merged into the
Turkey!

The *Bronzed* or American sub-variety appears
really to be composed of some immediate and
domesticated descendants of the wild American
race ; and which are distinguishable from the
common breeds by plumage of a dead black, but
so radiant with metallic reflections of the most
brilliant hues, that they are said to be quite " *dazzling*, when seen in the sunshine ;" although, we
confess, we have never detected such " dazzling "
refulgence, it is yet undoubtedly the fact, that
fowls of this breed possess a peculiar and very
beautiful bronzed appearance in their plumage,
unequalled by any other kind in this country.
They also attain a greater size than the common.

Black.—The Rev. E. S. Dixon divides this sub-
variety into two, which he calls Norfolk and Cam-

bridge,—the former being (he states) of a dull
rusty black, with brown tips to the feathers, whilst
the plumage of the latter is described as shiny
black with a bluish tinge,—the size also being
larger. In this division we cannot concur; we
have not ourselves been able to detect this difference in the blacks reared in those counties; and
as to size, we believe either may go into the scale
against the other. It is perfectly true that in *black*
Turkeys two shades are often seen, the one a bright
lustrous black, and the other a dead black waved
with very rusty shades, especially upon the back
and tail coverts; but the two shades are equally
found in the so-called Cambridge breed as in the
old Norfolk,—both indeed form but one sub-variety. May not the superior lustre or brilliancy
in the plumage, which some blacks undoubtedly
possess over others, indicate merely the intermixture of the fresh American blood, which some years
ago was introduced into this district, for the purpose of preventing degeneracy; and would not this
also account for any superiority of size such fowls
might enjoy?

GUINEA-FOWLS.

Sub-genus Numida.

Name and History.—The Guinea-fowl, or Pin-
tado, (as it is sometimes called,) is described by

several ancient Greek writers; and it also bears its part in their fabulous narratives. As we have already stated in our account of the Turkey, the generic name of that fowl has been stolen from the Guinea-fowl,—the latter being the true *Meleagris* of the ancients. According to Mythology, the Meleagrides, the sisters of Meleager, (son of Œneus, king of Macedonia,) who was cruelly put to death, bewailing the death of their unfortunate brother, were metamorphosed into Guinea-fowls, the showers of tears they shed bedecking their otherwise sable plumage with white spots; and were it not too *grave* a subject to joke upon, we should say that this melancholy origin may account for the peculiarly mournful cry of " Come-back!" which the Guinea-hen utters to this day—that invocation being no doubt addressed to the departed Meleager! By old Latin writers they are styled *Gallina Africana*, and Numidian Fowl. In Portuguese they are called *Pintada*, and in Spanish *Pentates*,—both words signifying painted, a not altogether inapt embodiment of the idea which their regularly marked feathers convey. After the allusions made to them in classic authors, they appear to have been lost sight of until their introduction into America by the Spaniards, in the early part of the 16th century, with the first cargo of slaves from the shores of Africa. In this country they were probably received subsequently from the coast of Guinea. Among us, in addition to the names already given, they are known as " Pearly Fowls," in allusion to the pearl-like dots upon the plumage; and in Ireland they are popularly known as " Gallins," evidently a corruption from the Latin *gallina*. Originally, as we have said, natives of Africa, the Guinea-fowls are now common in most parts of America; they have also been conveyed to Jamaica and other West Indian isles, where, however, they are said to remain perfectly undomesticated. They are distributed over most of the countries of Europe, except the more northern : they appear to thrive very well (though but little cared for) in England, but are scarce in our sister countries. They are found plentifully along the shores of the Mediterranean; and in the East Indies are much kept in a domesticated state by the Europeans, who carried them out with them. In their primitive habitat they are met with in large flocks like the Turkey; chiefly frequenting the banks of rivers and the humid localities of close herbage and brushwood.

COMMON GUINEA-FOWL. (*N. Meleagris.*)

Description.—This fowl is about the size of a good Dorking hen, although many place its weight at from 3 to 4 pounds only—certainly much below the actual fact. It is nearly 2 feet long, but, from standing very high upon the leg, and owing to the very loose and full plumage, it has the *appearance* of being a much larger bird than it really is; it is, however, of a round and compact form, having the back very much arched.   The plumage is of a dusky black, with a sort of bluish or dark slaty tinge throughout; each feather is uniformly dotted over with small circular white spots, varying in size in different parts of the fowls,—the feathers on the wings, tail, and under portions of the body being usually covered with larger spots than the rest; the flight feathers of the wings are mostly longer than in other poultry, though some Guinea-fowls possess much stronger powers of flight than others: the head and neck feathers have peculiarly soft, downy, loose webs; the former is surmounted by a short, thick, horny excrescence, generally called a " casque," from its resemblance in shape to a helmet, beneath which, over the bill, is a livid fleshy membrane; from the lower portion of the bill depend some broad flat wattles of a bright red colour; under the eye and upon the cheek is a livid white naked patch; and above the eye is a very arched and strongly-marked line or eye-brow, giving the fowl a most grotesque and ridiculously astonished visage; legs smooth, without spurs, and of a dusky straw colour; tail very short, and turned downwards like that of the partridge tribe. —The female, or Guinea-hen, is so nearly like the male, that it requires far more than ordinary attention and observation to distinguish the sex,— especially as the cocks are spur-*less*.   The principal points in which a difference (however slight) may be with care discriminated are these : in the hen the proportions are somewhat smaller, the frontal casque does not extend quite so far backwards, nor so high, and the wattles are smaller and frequently of a leaden hue compared with the florid red colour of the male. In the very peculiar and monotonous cry of these fowls, the Rev. E. Dixon states there is another and unerring distinction, that *besides* the shrill, harsh call-note *common to both*, the *hen alone* utters the mournful cry of " Cum-bec! Cum-bec!" which has obtained for them both the provincial name, in Norfolk, of *Come-backs,*—but which, in another sense, they are little entitled to, when they chance to stray away from home. (*Plate* V. *fig.* 2.)

Sub-varieties.—No tribe of fowls presents less difference or variety among its members than does the Guinea-fowl: and although it has been asserted by many that there are " six distinct species," (by which, it is presumed, *varieties* are meant,) there are certainly not more than two such,—the Common, above described, and the Tufted or Crested. There are, however, some which exhibit certain *variations in the plumage,* which may be appropriately noticed as sub-varieties of the Common Guinea-fowl; these are,—

WHITE, or Albinos (called "Mitred" sometimes). | MOTTLED, or Pied.
GREY. | LACED, or Black-Spangled.
| BLACK, or Negro.

The *White* (so called) has ground plumage of

a dull creamy shade, speckled over with indistinct and not very uniform spots of a clear white : they are almost exclusively confined to the north of England, and even there are by no means common, and still less handsome.—The *Grey* differs only from the common type, in having the ground colour of the feathers of a bluish grey, or dull lavender shade, instead of black.—The *Mottled* have the breast white, and also the primary or flight feathers of the wing; but in every other point, as to form and colour, they are identical with the original type,—indeed it is a question with us, if they are not merely *imperfect* examples of the primitive stock.—The *Laced* are of exceedingly rare occurrence : in them the colours and marking are entirely reversed,—the ground feathers are white, and the dots black, producing a very singular effect,—not much unlike a Silver-laced Bantam.—The *Black* is of an intense black colour, with the dots or spots *almost entirely obliterated*.   Our remark in reference to the origin of the Mottled applies equally to this breed.

TUFTED OR CRESTED GUINEA-FOWL.
(*N. cristata.*)

This bird is said by naturalists to inhabit Sierra Leone and the great Namaquas country. It is smaller than the Common variety, from which it is unquestionably distinct; though in form it closely resembles it. The head and neck is covered only with a loose naked membraneous skin, very like the hood of a Turkey, and of a similar livid hue; from the forehead, and over the crown, spring some feathers with disunited webs, forming a black tuft or crest in place of the horny casque or helmet which distinguishes the Common Guinea-fowl. The feathers are spotted over with small greyish dots, which vary from four to six in each feather, and are encircled or edged round with an intermediate shade to the ground plumage; the primaries of the wing are dusky brown, the secondaries being edged with white.

Some mention an "*Egyptian*" variety of this fowl; of which, however, they seem to know nothing except that it has "double wattles."

PIGEONS.

Gen. Columbidæ.   Sub-gen. Columba.

THE Pigeon is recorded as one of the most ancient inhabitants of the feathered world. It is found wild in most of the countries lying between the warm and temperate zones,—seemingly flourishing equally well beneath the scorching tropical suns, and in the mild climate of temperate regions, and avoiding only the immediate vicinity of the poles. From their exceeding beauty and innocence, they have ever ranked among the chief feathered favourites of mankind; in the countries of the East (the source of many of our superstitions) the Dove or Pigeon has always been a great and especial object of veneration; from the earliest ages it has been regarded as the symbol or emblem of numerous Christian virtues and graces, which, if not altogether " on earth unseen," are supposed to appertain in a peculiar degree to the character of this bird. Among some of the nations of antiquity it was held sacred, and no one dared to assail its life.

Among the wild race, modern naturalists enumerate a very great number of varieties and kinds; of these, however, only the following five are known to be inhabitants of Europe, and to them our notices of the undomesticated Pigeon will be confined :—

STOCK DOVE, or Wood and Bush Dove.

RING DOVE, or Cushat, and Wood Pigeon.

TURTLE DOVE, or Ring-necked Turtle.

COLLARED TURTLE, or Laughing Dove.

ROCK DOVE, or Blue-Rock Pigeon.

The STOCK DOVE (*C. Œnas*) is larger than the common dove-cote Pigeon : the plumage of a bluish-grey or lavender shade, varying from the darker tints to that peculiar hue of light or " French " grey, (the characteristic prevalence of which, in the feathers of birds of this tribe, sometimes obtain for it the designation of *dove-colour*,) with a dull reddish tinge pervading throughout : the sides of the neck of a rich bronze colour, and the breast darkened with fine changeable reflections of green and purple. It inhabits principally mountainous and wooded districts : it would seem to be easy of domestication, so far at least as attempts have been made ; but it is by no means common in this country, being seldom met with, except in the midland counties,—although it has been known to take up its abode sometimes in the warrens of the open counties of Norfolk and Suffolk. According to Buffon, (whose systematic plan of accounting for the numerous varieties in the different races of animals is far more convenient than correct,) the whole Pigeon family is derived from one common root—namely, the Stock Dove, an erroneous opinion long since exploded.

The RING DOVE (*C. Palumbus*) is the largest of all our European wild Pigeons. Its general appearance and plumage much resembles that of the Stock Dove, but the feathers are darker and more of a leaden grey; it is chiefly distinguished by several broken white rings which nearly surround the neck; and the coverts are edged with white likewise. It seems scarcely capable of domestication : like the Stock Dove, it dwells in the wooded parts of Great Britain, but is also found very plentiful in Germany. They have a remarkable note of singularly sad and mournful cadence, for which some of their names have been bestowed upon them.

The TURTLE DOVE (*C. Turtur*) is but little more than half the size of the last-named. The general colour of the plumage is a dull slaty-blue, shaded with light, dark, and reddish brown; the bottom of the sides of the neck being, like the Cushat, almost encircled with three or four streaks or bars, but in this case they are black tipped with white. It is distributed in most of the temperate countries of Europe. Among us it seems also tolerably well naturalized, being found alike common in the eastern, midland, and western counties of England; and have even strayed into the northern parts of Britain: they are, however, only summer visitants, and, as winter approaches, seek more genial abodes in southern climes. On the domain of Warwick Castle, there were, some few years since, (and, for aught we know, may be still,) a greater number of Turtle Doves than could be met with in any other district here: they abounded in multitudes throughout the woods and plantations, the turrets of the castle being also a favourite resort for them, whence their rather melancholy cooing was heard to a considerable distance along the road.

The COLLARED TURTLE (*C. Risoria*) is rather larger than the preceding. The plumage is of a reddish-white or dull salmon colour; the back of the neck marked with a thick black streak, which tapers off round the sides, forming a tippet or collar, the points of which scarcely meet upon the throat. It has a peculiar note or cry, which bears (it is said) some faint resemblance to the sound of laughter, and hence its second name. The original habitat of this bird, there seems reason to believe, must be in India and China, but it has long been known over Europe, and it is also common in Syria and countries adjoining. The ease with which it is domesticated, it is alleged, confirms the opinion naturalists entertain, that this is the pigeon so frequently alluded to in Holy Writ.

The ROCK PIGEON. (*C. Livia.*) If, as some have thought, the repeated allusion to pigeons which occur in the Sacred Volume refer to the Collared Turtle, it is at least equally certain that the Rock Pigeon was familiarly known to the ancients: thus the prophet Jeremiah, apostrophizing the " dwellers in Moab," exhorts them to " leave the cities and *dwell in the rock like the dove,* that maketh her nest in the sides of the hole's mouth,"—a more direct allusion than which cannot perhaps be found. It is this wild variety also which many writers have assumed to be the common origin of most, if not all, of our domestic breeds; the correctness of this opinion, however, (notwithstanding the sanction given to it by so high an authority as Mr. Yarrell,) we fancy admits of a doubt. This variety is extensively distributed in various parts of the world, inhabiting, as its name would indicate, the clefts and crevices of rocks and caverns upon the sea-coast. In our own country they abound upon the steep and rocky portions of the southern coasts; and are not unfrequent visitants to the eastern shores, occasionally alighting on the beach at Great Yarmouth,—although the flat sands there afford no abiding places for the birds, which thence wing their flight often to a considerable distance inland : to the north, the dangerous rocks and reefs at Flamborough Head have ever been favourite resorts for them ; and they are also met with along the eastern coast of Scotland, the Orkney and Shetland Isles, away as far north even as Norway and the Faroe Islands. The Rock Pigeon is not larger than the Turtle Dove: the general colour of its plumage is a fine bluish or French grey ; the feathers of the tail and some other parts shaded or tipped with a darker slaty lead colour ; the wings barred across with two bands of bright black: the lower part of the back at the base of the tail runs into a patch of silvery white,—a characteristic especially to be observed in this variety, and which distinguishes it from other wild races.

DOMESTICATED VARIETIES.

Of this class Buffon enumerated, in his usual loose way, upwards of thirty " varieties," which some writers have swelled to seventy, and others to a hundred! Descended, as they assert, from one common stock, all the variations of form and colour, which we see, are attributed to human

ingenuity and contrivance, in the skilful exercise
of fancy. It seems to us, that there exists, nevertheless, some very essential differences in these
birds, which, however much they may or may not
be attributable to the nature of the region, soil, or
climate to which they are indigenous, are certainly
not the creation of the art or ingenuity of man.
And when it is said that the Rock Pigeon is the
original of our " Dove-cote pigeons," the phrase is
about as ambiguous as the hackneyed term " Barn-
door fowl." In the domestic race of *columbæ* are
comprised, we venture to hazard an opinion, very
many and *distinct* species, or (more properly
speaking) several *sub-genera :* but to enter upon
the discussion of the question of the *origin* of these
domesticated members of the pigeon family would
be perfectly beyond the limits of this volume, and
equally beside our purpose and design.

In this country the variety most usually associated with man is the common *Blue Dove-house
Pigeon*, (vulgarly " Duv-hus," or " Duffer,") as
it is called, which certainly bears a strong resemblance to its supposed wild progenitor, both in
form, size, and plumage ; the latter, however,
presents a darker and blotchy appearance : it is
distinguished from the Wild Rock Pigeon by
being mottled with several white spots upon the
wings and coverts,—marks which are *invariably
absent* in the other. Notwithstanding that these
pigeons are to be found in almost every part of

England, our inquiries lead us to believe that
pure specimens are far less common than they
were some years since; whether this is owing to
their being really only a semi-domesticated variety
reclaimed from their primitive state, or that they
have been neglected by attention being directed to
the rearing of fancy breeds, or that they have been
gradually drafted off into other and mongrel
breeds, we know not, but possibly all these causes
may have contributed to produce the result.

As we have already said, from the most remote
antiquity this beautiful and variegated race of
birds has been cherished by man as a source of
profit, and as providing an article of provision for
the table; whilst among the moderns they have
been scarcely less prized as a source of amusement, and of gratification to the eye.   At the present day, those varieties which are kept only for
amusement and show, are styled " Fancy Breeds,"
and form a distinct article of commerce in cities
and great towns,—the different sub-varieties, as
they chance to be alternately in fashion, sometimes
bringing considerable prices.

In London the pigeon-fanciers have for a very
long period, we believe, had a club in which pre-
miums were awarded for meritorious specimens
of birds, and the notable science of " The Fancy "
is promoted and perpetuated through the medium
of breeding and crossing colours and forms.   Half
a century or more ago the pigeon fancy was in

very much higher estimation and prosperity than at present; this decline may in some degree be occasioned by the practices of many disreputable persons who have resorted to pigeon-keeping, not from any real or harmless love of the Fancy, but simply as affording them the means and scope for the successful exercise of their low and thievish propensities: from the recent impetus and encouragement given to poultry rearing, it may, however, be hoped that pigeons equally with fowls generally will partake of the improvement which is manifesting itself in the study of that branch of rural economy, and that the Fancy will again revive, tending as well to the extension of our scientific knowledge, as to the promotion of innocent recreation and enjoyment.

Before entering upon a description of the different varieties and sub-varieties of the domestic race, we are anxious to disarm too severe criticism, by candidly apprizing the reader that we lay no claim to any greater practical or original knowledge than can be attained by keeping pigeons upon a scale by no means extensive,—confessing that we have never had the honour to be initiated into the histories or mysteries of the " Fancy;" on the latter part of the art, the best authority with which we are acquainted is a scarce and anonymous *Treatise on Domestic Pigeons*, published by C. Barry, Fenchurch Street, containing some very fair plates of fancy pigeons, but which is almost wholly compiled and drawn from a still older work, entitled *The Columbarium*, by John Moore, 1735, —this last is, however, quite inaccessible from its rarity. We also cheerfully acknowledge our obligations to a *Treatise on Pigeons*, by J. M. Eaton, 1852, which we recommend to those who desire further information, as being decidedly the most complete book that can now be had on the subject; it contains a reprint of Moore's Treatise, and though somewhat overlaid with irrelevant and not always quite intelligent remarks, is yet withal quaint and amusing,—its principal value, too, consists in the originality and practically useful information scattered through its pages. In regard to classification, the different members of the pigeon family are even in a more scattered and confused condition (if indeed that is possible) than the ordinary domestic fowls,—no attempt having yet been made with the view of classifying or arranging the several varieties, according to their apparent relative affinities, or natural habits and instincts. After putting forth such modest pretensions to an acquaintance with " the Fancy," it will not, we trust, be thought that we assume too much in suggesting some temporary arrangement or classification of so numerous a section of the feathered tribe; and for this purpose it would seem convenient to divide them into three classes, —under the first of which we should include all such pigeons, as whilst possessing considerable ordinary powers of flight, yet seem, in common with other members of the Rasorial group, to be destined by their heavy structure and natural inclinations and habits to live upon the ground, and which might therefore be not inappropriately termed *Columbæ Terrestrium*, — in the second would be ranged those pigeons having extraordinary power of wing, with which they cleave the air, and soaring upwards, love to pass the principal portion of their lives in worlds of clouds, such being distinguished as *Columbæ Cœlorum*,—and in the third division might be placed such pigeons as exhibit no decided indication of belonging to either of the preceding classes, but which possess some characteristic peculiarity of plumage, either by having a tuft of feathers projecting and reversed upon the head, or else encircling the neck. These classes would·of course embrace all the divisions into varieties and sub-varieties in which the *Columbæ* are comprised, and which may be reduced into the following tabular form.

I. *Columbæ Terrestrium.*

RUNTS.—Leghorn and Roman Runt, Spanish Runt, Egyptian Runt, Laughing Runt; Archangel, Porcelain, Magpie, Spotted; Frilled-back, or Frizzled; and Silky Laced.

FANTAILS, or SHAKERS.—Broad-tailed; Narrow-tailed.

CROPPERS.—Dutch, Uploper, Parisian Pouter, English Pouter, and Pouting Horseman.

II. *Columbæ Cælorum.*

CARRIERS. — Horseman, Scanderoon, English Carrier, Antwerp Carrier, and Dragoon.

TUMBLERS.—Common or English, Mottled, Ermine or Almond, Beards, Bald-pates, Helmets, Smiter or Drager, and Skinnums.

III. *Columbæ Cristatæ.*

Trumpeter, Finikin, Turner.
Barbe or Barbary, Mawmet or Mahomet.
Turbit or Owl, Nun.
Ruff, Jacobine, Capuchin or Carmelite.

RUNTS.—Of the whole pigeon race the Runt is the largest by far; in size and weight it equals and often surpasses the good average of the Bantam tribe. In point of antiquity also they are not inferior to any other breed: they are noticed in the writings of Pliny and other old naturalists. They are very common throughout the whole of the Roman states and Italy, and abound every where along the shores of the Mediterranean, whence, indeed, they appear to have reached us by the Levant traders, to whom, as we have already seen, we are indebted for other descriptions of poultry. The Runt is of a remarkably round, plump body, with broad back, and large fully-developed breast; neck thick, and large head of goose-like form; the legs, in comparison with the size of its body, are seemingly short and thick; the bill, which is strong, but of moderate size, has at the upper base a full wattle, (as it is popularly, though incorrectly, termed by Fanciers,) of a puffy, fungus-like excrescence.   A peculiarity in this variety is the manner in which the tail is carried, namely, like that of the duck, horizontal, or even slightly pointing upwards, instead of bending downwards as in others of its species.   The Runt is sometimes called Turkish and Russian,—a confusion of names, perhaps only owing to their having been brought from the Levant, in vessels which on their voyage may have touched at some port in either of the above-named countries.   The origin of their name seems scarcely explicable : among old Lexicographers, *Runt* has various meanings attached to it,—thus in English it denotes a three years' old Canary, a sour, crusty fellow; and among the Scots it is applied to an old cow,—none of which bear even a distant reference to the pigeons of which we are speaking ; but it seems to us, however, that a far more probable derivation may be the Belgic word *Rund*, signifying a short fellow, in allusion, perhaps, to their short-legged and squat appearance—an origin by no means unlikely, when it is remembered that formerly the Belgæ were the principal rearers of pigeons in Europe ; although, in the sense used, the name might have been bestowed with even more appropriateness to other varieties.

1. *Leghorn Runt.*—Generally considered the *best* of its kind, which, however, as a class are held in less esteem than their economical value would

appear to deserve.   It is found of various shades of colour in the plumage,—brown, tawny, mottled, light and dark slate, and even white ; it attains to an almost incredible weight for a pigeon, single birds having been known to reach 2½ lbs. : the flesh and feathering is said to be firmer and closer than that of other Runts.—*Roman Runt*. Though mentioned by some writers as a separate sub-variety, we cannot discover that they differ in any way from the kind previously described, which are met with throughout the Italian states.

2. *Spanish Runt*.—Described as the *largest* of the Runts : is much longer-bodied than the Leghorn, in some instances measuring more than 20 inches between the extremities of the bill and tail ; it stands shorter upon the legs, and is altogether of a broader, more squat appearance than the first, whose size it also exceeds.   The plumage is said to be of " all colours," although French writers speak of the Spanish or " Andalusian " Runt, as being *invariably* of a black or dun shade. A Spanish gentleman some time since assured the author that in Barbary there are Runts of the size of ordinary domestic fowls, but which, from their extraordinary weight, are incapable of flight.

3. *Egyptian Runt*.—This is much smaller than the preceding, though still of large proportions : they are not unfrequently brought from Alexandria by sailors voyaging thither.   We have never seen them but with pure white plumage, or of a

tawny cinnamon shade. These Runts are always feather-legged, (as are, very rarely, those previously described,) and sometimes slightly tufted upon the head.

4. *Laughing Runt.*—These appear to be evidently Runts; they are of middling size, and said to be brought from the Holy Land, or Syria. They are of two shades only—dark slaty-blue, or tawny-red mottled, but more generally the latter. Their only peculiarity would appear to consist in the coo, which is described as resembling the gurgling or bubbling of water, and "soft laughter," —thus hinting, perhaps, at some distant affinity with the Turtle Dove.

5. *Archangel.* — We place this sub-variety in the same category with the Runts, to which race it certainly seems to us allied; and although not of such large proportions, it is still above the average size of pigeons generally. In the head and bill the Archangel is very Runt-like, and altogether its form may be said to be inelegant; it seems probable, therefore, that but for the uncommon and striking colours of the plumage, the breed would never even have attracted ordinary notice. The colours are two—a rich bright chestnut, or metallic bronze or copper hue, and a lustrous blue-black, softly shaded and blended into each other; the former shade extending over the head, neck, breast, and upper portion of the back, whilst the wings, tail, and rest of the body is the beauti-

ful black we have said.   Occasionally, a short tuft
or " turn " of feathers makes its appearance at the
back of the head ; but specimens being as frequently met with in which this tuft is entirely absent, we
do not regard it as any *characteristic* of the Arch-
angels.   This breed was first introduced (perhaps
from the place whose name it bears) at Knowsley,
but they are still uncommon, which may be attributable to their being very little prized or
sought after.

6. *Porcelain.*—The only qualification of this
nondescript (apparently *accidental*) breed is the
singular pied plumage,—though even in this particular different specimens exhibit but little characteristic identity: some are black mottled (though
in a different way to the Tumbler) with white,
and have certainly a strikingly pretty appearance
as a " Toy " bird: others have the head, neck,
and crop part a dark brown or black, and the
back and body quite white, with a tolerably broad
reddish sort of salmon streak across the lower or
saddle portion of the back, with a narrower of the
same shade running into and ending with black.
The name was probably given to it from the fancied resemblance of the peculiar mottling or pie-
ing of the plumage to the absurd colouring often
seen in the familiar chimney ornaments of china,
(or *porcelain,*) representing—or intending to—
cats, dogs, &c.

7. *Magpie.*—This is probably of the same class

as the preceding one; and is described with plumage much like that of the magpie.

8. *Spotted.*—About the smallest of the Runts. The general plumage is white, with a peculiar "spot" of another colour (sometimes red, black, yellow, or slate) upon the forehead, and occasionally upon the wings. The tail is of the same shade as the spot, whatever colour that may be.

9. *Frilled-back*, or *Frizzled.*—In shape it is very much like a Runt, though inferior in size. The plumage is invariably white—at least all the specimens that have yet made their appearance (and those have been but few) at the Poultry Shows, have been of that colour: the peculiar and distinguishing characteristic, whence it takes its name, is the singular appearance of the feathers, which seem throughout as if *curled* or *frizzled* upwards at the ends, in a manner very similar to that we have described in our account of the "Frizzled Fowl," at page 188.—The *Frizzled*, or (as it is often erroneously called) the "Friesland" Pigeon, we regard as identical with the Frilled-back,— certainly we have never met with a fancier who could tell us the difference (if any) between them; and it seems to us that it is only in the *name*,— the first being used by old writers, and the last by gentlemen of the modern fancy, to indicate the same peculiarly feathered variety. They are said to be feather-legged occasionally.

10. *Silky Laced.*—Of this singular variety we

believe the only specimens seen in this country are those in her Majesty's collection at Windsor, though it is described as common in Holland, being " originally bred there." In size it would seem to be the smallest of the Runt family. The plumage is entirely white, and the feathers throughout present the same remarkable anomaly of having the webs divided into silken filaments, in a way similar to that described in the " Silky Chinese Fowl," at page 190,—indeed, it seems to occupy the same position among pigeons as that fowl does among common poultry.

11. FANTAIL, or SHAKER.—This singular and very pretty race is characterized by an extraordinary number of tail feathers, which (in proportion to the size of the bird) are also developed in a much greater degree than in other varieties,—the usual number of such being *twelve*, whilst in the Fantail they are found to be twenty-six, thirty, thirty-two, and even (though very rarely) as many as thirty-six: and these feathers add further to its singular appearance, by being always, except in flight, kept expanded like the coverts of the turkey, and carried much elevated upon the back and drawn inwards, gradually sloping to the back of the head. In size it is not large; head small and round; neck long and very slender, and bent or thrown backwards until it nearly meets the tail, the breast being protruded forward, and the wing dropped down below the body,—affording some striking

points of resemblance to the Sebright Bantam. The plumage is generally throughout of a beautiful snow-white; though specimens occasionally occur perfectly black, and more rarely still, pied or mottled upon the head and shoulders with a slaty-blue, or a dull reddish colour, like Turbits. They are called *Fantails*, from the peculiar slope and expansion of their tail,—and *Shakers*, in consequence of a remarkable tremulous motion, or quivering tremor, which agitates them when excited by anything. (*Plate* VII. *fig.* 3.)

12. *Narrow-tailed Shaker.* — Although some fanciers hold to the opinion that this is distinct, we are inclined to believe it is only a bastard strain from the true Fantail, from which it differs but in the shortness and thickness of its neck, and the much reduced proportions and number of its tail feathers, which though expanded are carried only very slightly elevated : whereas the points of excellence to aim at in Fantails are long thin necks, with broad, ample, and elevated tails. There is also this further presumption against their being genuine, that specimens, which would be called perfect *narrow-tailed* Shakers, have been produced from the true Fantail and some other variety.

CROPPERS.—This peculiar and isolated race appears to have been originally bred in Holland, but undoubtedly it has been brought to perfection in our own country. It is distinguished by, and receives its name from, an enormous *crop* or bag, which it has the power in a remarkable degree of inflating or distending with wind, until a sort of balloon-like globe is raised up round the throat, in which the head and neck seem completely buried. To counterbalance this addition in front, the Cropper carries itself very upright, the head and legs being almost in a perpendicular line. By this singular quality, its grotesque attitudes, and its familiarity with man, the Cropper engages attention, and attracts the curiosity of fanciers, with whom it has always been a great favourite.

13. *Dutch Cropper.*—This is the variety whence, by careful selection and breeding, the *Pouters* have been produced, and it is in every way inferior to them. The body is thick, short, and lumpy; the crop is not so capacious as in the Pouter, and generally sags or hangs down below the beak instead of enveloping the head as it were; the legs too are short, thick, and straddling; the thighs and legs are feathered quite down, but the feathers are loose, coarse, and quilly. The colour of the plumage is in almost endless variety.

14. *Uploper.*—So called from the Dutch word *uplopen*, (jumping or leaping up,) used to describe a habit, which it is said to possess, of leaping up and expanding its tail (though in a *reverse* position to that of the Fantail) when courting. It differs but little from the preceding, but is somewhat smaller, and the legs are set closer together and more slender; the toes are short, and diverge very

little from each other. Moore asserts that they are scarce, and generally found of one colour—white, blue, or black, and not pied or mottled, like other Croppers.

15. *Parisian Pouter.*—This bird, which was formerly known by the vulgar name of "Parazence Pouter," is described as having "all the nature of a Pouter, (or Cropper,) but is generally long-cropped and not very large:" probably it does not merit the title of being a distinct breed from the old Dutch Cropper. It was chiefly ad. mired for a very beautifully variegated plumage, " being chequered with various colours in every feather," but we have not seen any notice of it by recent French naturalists.

16. *English Pouter.* (*Plate* VII. *fig.* 2.)—Here the " Cropper " fancy seems to have attained the summit of perfection in the breed,—all the points most admired having been brought prominently out. We shall perhaps best describe the beau-ideal of an English Pouter by giving the following points of excellence, as established and laid down by the gentlemen of the Fancy, and arranged in the order their importance seems to us to render necessary.

   a. Size.—In this point it may be said that the English Pouter is a *tall* and *slender* bird, rather than *large* or *stout:* generally amateurs do not sufficiently discriminate this distinction and difference,— and as a conse quence, they breed them much too unwieldy and of almost Runtish proportions. Length, measured over the head, from the point of the bill to tip of tail, from 17 to 20 inches; and height, from 13 to 15 inches from the ground: length of leg, (measured from upper joint of thigh to *commencement* of the toe-nail,) from 6 to 7 inches: the thighs and shanks being of moderate thickness, neither wiry nor clumsily stout.

b. Shape.—Slenderness of girth is a very great point, the measure as much as possible within 7½ inches, it gives a sleek and smart appearance to the bird: the back should be hollowed out, tapering from the shoulders, and sweeping gently down to the tail,—though this depends very much upon the possession of the next quality.

c. Carriage.—No unimportant consideration in a Pouter: it should be so erect and upright that the head be carried almost in a perpendicular line with the legs: and when playing, the tail should be spread somewhat like a fan, but not so as to " scrape the ground." The wings should be carried close to the body, and the legs set near to each other, (not straddling,) which gives the bird a sort of tripping or mincing gait when walking.

d. Crop.—This should be large and round—in fact, a perfect globe; carried well up under the bill, and filling all round behind the neck so as to bury the shoulders, as it were. It should be neither too distended with wind (termed "buffled,") nor "slack-winded,"— for in the former case it is so strained or inflated that the bird is seriously incommoded, and gives the appearance of very scanty plumage, much naked skin being exposed; whilst in the latter case, the crop hangs loosely, and the chief characteristic is lost.

e. Feather.—As to this last, the colour of the plumage must be simply a matter of opinion and taste among fanciers: the pure white are rather scarce and handsome; but the pied kinds are the most admired, and are found of black, blue, red, and yellow (so-called) mixtures; in any case the head, neck, and back should be of a uniform whole colour, the white being shown upon the front of the crop, just below the throat, not however running quite round, but forming a sort of horseshoe belt, termed the "bib;" the flight feathers of the wings also all white; and the legs from the thighs down to the toes should be covered with close all white feathers, fine, soft, and downy: the tails of the black and blue pied are of the same respective colours, whilst the red and yellow pied have *white* feathered tails. It may be noted that the blue pied should have two streaks or bars of bright *black* across each wing. There is likewise a sort of fawn shade (known provincially as " cloth" colour) which is common, but will not compare, we fancy, with any of the other kinds.

17. *Pouting Horseman.*—Stated to be an unquestionable cross, bred between the two varieties whose names its own is compounded of. It differs little from the Pouter, save in being much slenderer in the girth, and of smaller proportions generally. It may be said to be a reduced facsimile of the English Pouter; and the two have been not inaptly (though fancifully) described by comparing the latter to a fine portly old English gentleman, while our subject personifies in its smart, stylish person, the fast young English gentleman. The admixture of the Horseman's blood gives this bastard very considerable power on the wing, and it therefore appropriately closes this class, and leads us to a consideration of the next.

II. *Columbæ Cælorum.*

CARRIERS.—The useful qualification of *messenger* or *carrier,* appertaining to the Asiatic and African varieties of the pigeon, is of high antiquity; and from them doubtless has proceeded the breed known in Europe under the name of carrier. During the crusades we read of an

Eastern prince who had a sort of telegraphic communication constantly kept up in his dominions through the medium of trained pigeons, relays of which were placed at regularly appointed posts, in constant readiness. In recent times they have been employed in a similar way in this and in other countries, as expresses for conveying early intelligence on commercial or sporting matters, and of events of interest or importance, although the telegraph has now superseded them. As might be inferred, the Carrier tribe possess a prodigious power of flight, far surpassing that of any other variety, and with greater powers of sustaining it; the wings are large, powerful, and rigid, and supported upon strongly knit shoulders and by a broad muscular development of the breast, by which they are enabled to cleave through the air with almost incredible swiftness : the rate at which they have been known to fly has been stated at from thirty to one hundred miles per hour. The greatest authenticated distance accomplished on one occasion by any of these birds is about six hundred miles, namely, from St. Sebastian in Spain to Vervier. It is recorded of a merchant at Aleppo, who for a large bet flew one of his Carrier pigeons, in the hope of its reaching Alex. andretta within the prescribed time; but mounting very high, it must have mistaken the sea of Balsora for the Mediterranean, and winging its flight thither, missed its destination and passed over to the Indies, whence it returned in three days; the distance it must have traversed being proved by the owner, who, enraged by the loss of his bet killed the pigeon, when its crop was found filled with green cloves, which could only have been procured from the quarter indicated. The *rationale* of pigeon-flying would afford matter for the most interesting consideration, and for ingenious speculation—but our limits preclude our entering upon it.

18. *Horseman.*—Now generally considered as the original type of the Carrier group; though formerly supposed to be a bastard strain. It is a fine large bird, having a long body, head, and neck; bill long and stout, and covered at the nostrils by a tough leathery substance of a white fungous nature, which is often improperly termed "the wattle;" eyes reddish, and encircled by a broad band or rosette of a similar nature to that over the bill: an ample breast and broad back and shoulders, added to its being set upon high strong legs, give the Horseman the appearance of a pigeon of more than ordinary proportions.— *Scanderoon.* This we believe to be only a different name for the Carrier or Horseman of the East, which is bred extensively at Scanderoon, (also called Alexandretta,) a seaport in Syria, whence formerly large numbers were kept in order to be despatched with intelligence of the arrival of vessels there: this breed is styled by French writers

" the Swan-necked Egyptian." The Horseman is found in almost every variety of colour—white, reddish, blue, black, and even pied or mottled.

19. *English Carrier.* (*Plate* VII. *fig.* 1.)—This stands in the same relation to its kind as the English Pouter does to the Cropper—its superiority consists only in the *perfectness* to which all the *points most admired* have been brought after long and careful breeding. In the grace and general symmetry of its form it is unsurpassed (perhaps unequalled) by any other variety; whilst its proportions also exceed those of pigeons generally,—the length from the tip of the bill to the tip of the tail often reaching in adult specimens sixteen inches; in height it has been known to attain nearly twelve inches—but little short of the Pouter: the weight of a single bird varies from 19 to 22 ounces. The points of excellence, or " properties," of the English Carrier we should arrange in the following order: and although exception may be taken by the Fancy to the prominence given to the second point, we think if, not alone regarding it as an object for admiration in the aviary, its peculiar nature and adaptation for the office of a winged messenger be also considered, the precedent importance of the means and powers of flight will be at once apparent.

a. Shape,—elongated; great breadth at the shoulders across the back, and slenderness of girth; ampleness of breast; neck long, slender, and only very slightly bent.

b. Powers of flight.—General development of strength in wing, and great length of the flight feathers.

c. Bill.—Straightness, thickness, and length, are the three requisites,—the two first the most indispensable in the eyes of true fanciers; and for the matter of the last, an inch and a half measured from the inner angle of the junction of the upper and lower parts of the bill to its tip, is considered a high standard: of a dark olive colour.

d. Wattle (so called).—Full, puffy, rising high above the bill, and broad upon the top: having an inclination forwards, and running over the sides until joining a similar but smaller excrescence upon either side of the lower bill: formerly this fungous skin was preferred when possessing a slightly blackened hue, as if gently dusted over with a dark powder, but this we believe is no longer insisted upon.

e. Head.—Elongated, narrow, and much flattened or depressed upon the top or crown.

f. Eye.—Full, bright, and most preferred of a fiery red colour: the fungous rosette surrounding it broad, round, thick, and of uniform width—the diameter of the whole circle being from 1 inch to 1¼ inch; the puffy substance rising even above the level of the top of the head. In the female this rosette and likewise the " wattle " is developed in a very much less degree.

g. Feather.—On this point the fancier must alone be guided by his taste : there are Carriers of various colours,—white, blue, dun, black, and (it is said) pied: they are mostly met with black, but those of what is termed a *dun* colour have long been held in the greatest estimation, and perhaps not altogether without reason,—as it has been observed that kind generally have very perfect heads and bills. Carriers of the darker shades of plumage have peculiarly brilliant metallic reflections upon the neck and breast feathers.

20. *Antwerp Carrier.* (*Plate* VII. *fig.* 1.)—This is comparatively speaking a recent introduction into this country from Belgium, having been originally bred at the city after which it is called. But very little was known of it until a noted Pigeon Match in July, 1830, when a number of these birds, which had been conveyed hither for the purpose, were " tossed " (as the flying of them is called in the technical phraseology of the fancy) from London Bridge: more than nine tenths of these reached their destination (Antwerp) in a few hours. Although possessing very

remarkable and superior powers of flight, they have never been great favourites in the Carrier Fancy, which we believe is mainly attributable to the paucity of really genuine birds that have found their way to this country,—the mongrels passed off as such not deserving the name of Carriers. The origin and distinctiveness of the Antwerp is a matter on which great difference of opinion exists : by some even they have been thought a cross between the English Carrier and the Owl pigeon and Dragoon, or some other variety. Mr. R. P. Brent states positively that they are descended from a small variety of Rock Pigeon (?) peculiar to the city of Antwerp and the surrounding locality. Whatever their source, their properties certainly render them a valuable acquisition, and deserving of greater attention than they have hitherto attracted. The *true* Antwerp is much inferior in size to the English Carrier, never exceeding 14 ounces in weight: the general proportions, too, are more sleek and slender, and consequently the contour is quite as graceful and symmetrical, though it may be less aristocratic or dignified in appearance : in length it seldom exceeds $14\frac{1}{2}$ inches, from the tip of the bill to that of the tail,—though here we may observe, that as the measure is taken *over* the head and back of the neck, the length of all pigeons depends much upon the mode adopted, considerable difference being frequently found arising from the voluntary elongation or contraction of the neck: the girth across the wings and shoulders will be found to be about 11 inches. In other points the Antwerp closely resembles the English Carrier, though possessed somewhat in a less degree: the head is not broad, and should be slightly depressed on the crown: the bill is broad and thick, not exceeding an inch in length at the gape: the skinny excrescence upon the bill and around the eye is only slightly developed, comparatively,—even less so than in the Horseman: it does not stand by any means so *high* as the English Carrier, chiefly because its carriage is much less upright. As a messenger we do not hesitate to say that the Antwerps are in no way inferior,—possessing equal powers of flight, and with considerable less weight to carry, they accomplish most incredible distances with great rapidity and less fatigue. In colour the Antwerps are usually of a dark dun and slaty-blue shade, at least the best specimens we have seen have been so: there is a mealy "cloth" feathered Antwerp, often barred with reddish brown across the wings, of larger proportions, and, we fancy, of a bastard strain.

21. *Dragoon.*—This breed (which is frequently vulgarly called the "Dragon") is held by the Fancy to be "absolutely and without dispute" a bastard strain, bred originally between a Horseman and a Tumbler,—the character of the former

greatly predominating, though smaller in size of body; as hybrids, however, they are rapidly being absorbed in the original Horseman, and although often presenting pleasing mixtures of feathering from the cross, their exclusion (in common with all other mongrels) from the Exhibitions is to be desired by all true Pigeon Fanciers.

TUMBLERS.—This variety is among the smallest and prettiest of the *columbæ*. The name is derived from the singular faculty it possesses, of "tumbling" when flying upwards, or rather of throwing a series of somersets forwards and backwards head over tail, and *vice versa*, especially when immediately mounting in flight or descending to the earth: it is however not alone for this peculiarity that the Tumbler is admired,—its singularly pretty shape and contour, neat and trim appearance, and the diversity and variegated nature of its plumage, have won for it a numerous circle of admiring fanciers. The Tumbler can have altered but very little since the publication of the *Treatise* more than a century ago, when it was correctly described according to the standard even of the present day: it is small and very compact in form; strong in the body; full in the breast—almost globular, and tapering gradually along the sides towards the tail, which is narrow; neck not long, but rather slender; head much rounded, with a remarkably full and round forehead projecting forwards; bill very short, thick, and what is termed "spindle" shaped; eyes most admired when the *irides* are of pearl-like whiteness; the legs and feet of singular (almost ridiculously) short proportions. The wing is remarkably rounded or bowed in the outline from the shoulder along the outer flight feather to the tip.

22. *Common* or *English.*—It seems to us that this cannot but be regarded as the primitive stock, whence the different sub-varieties which follow in course have from time to time been drafted off and established. They are generally coarser and larger in all points than the high-bred sorts,—being more particularly stouter in the body and longer in the bill: and it is worthy of note that the peculiar and amusing faculty of *tumbling* in flight is present in a much greater degree in the common kinds,—from which it may be inferred that the breeding of the Tumbler to the standard "points" is not accomplished altogether without the sacrifice of more pleasing and natural properties. In plumage they are usually of a whole colour, namely, all white or cream colour, light reddish-brown or cinnamon, and black, when they are vulgarly called "Kites:" but these shades are not unfrequently found splashed or "myrtled" (i. e. mottled) with white,—although the latter kind should perhaps be included under the following sub-variety. Mr. Morton Jones of the Crescent, Birmingham, has exhibited some speci-

mens of the Common Tumblers having tufts or
crests of feathers at the back of the head, after the
manner of the Turbit pigeons: but these he informs us, in a communication with which he kindly
favoured the editor, have been obtained, by crossing the Tumbler to the Turbit; and thus, by frequently pairing their young over to the Tumbler,
birds have been produced which, whilst retaining
the faculty of tumbling in flight, have the crest in
addition.   Old writers mention a "a pretty little
blue Tumbler with black bars" across the wings.

23. *Mottled.*—This sub-variety is, generally
speaking, more *carefully* bred than are the Kites;
and therefore more nearly approach the *beau ideal*
of the Fancy, whose principal aim appears to be
to produce the "*shortest-faced*" bird possible,
that is to say, a Tumbler with the largest or most
projecting forehead, and the smallest perceptible
appearance of a bill,—a property which is possessed in an extraordinary degree by the Almond
or Ermine to be afterwards described.   Of the
mottled, also, there were formerly every variety
of feather, but from greater attention being paid
to them, the chance variations of feather are less
frequent, and they are now bred with more uniform and regular markings.   The *White* mottled,
so-called, (but which should rather be described
as a *Pied* Tumbler,) is very seldom met with, and
by the unobservant has been confounded with the
true Black Mottled, from which, however, it

widely differs; its ground colour being white, with *black* flight and tail feathers, and others of similar colour scattered over the back and body in small patches: whereas the plumage correctly termed "mottled," is of a coloured ground (no matter what shade) mottled with a few *white* spangled feathers upon the shoulder of the wing, forming at that point a sort of mottled rosette, or as it is called by the Fancy—a "rose pinion," that being considered the most elegant: this white mottling is most admired when confined to the part indicated, although this is but seldom attained: and many contend that the true mottled Tumbler should also have a few white spangles upon the upper part of the back, as a relief to the uniform colouring of the rest of the body; a bird so mottled is said to have a "handkerchief back."—The *Yellow* Mottled is of a dull yellowish buff plumage, mottled in the way already stated.—The *Red* and the *Black* Mottled—the former with ground plumage of a dull reddish cinnamon colour, and the latter of a sable hue, the more intense the better—have their respective admirers.

24. *Almond* or *Ermine.*—At once the most delicate and beautiful of the Tumblers, this pretty little creature in a superior degree unites in its miniature compass all the points and properties most sought after and admired by the Tumbler Fancy,—indeed, it seems (if we may so express it) to be a sort of joint-stock production from the

preceding sub-varieties, the best birds of which have contributed that in which they excelled, (whether in form or feather,) in order to embody in one the highest attainable perfection of Tumbler breeding. But it is not alone for the elegance of its form, or the striking beauty of its plumage, that the Almond may be said to have taken the highest rank in the group,—it is the great scarcity of really good specimens, arising from the extraordinary difficulty experienced in breeding them at all conformable to the established standard, which has caused them to come to be regarded by the Fancy as almost the only variety exclusively worth their attention and pursuit; as such they are prized in a measure corresponding with their rarity. Their name is thought by some to be given them from the *ermine*-like mixture of white, yellow, and black, present upon the tri-coloured spangled feathering : although an opinion has long existed that the second name is its proper designation, as describing the peculiar *almond*-like shade of the ground-colour of the plumage so much sought after,—and we are inclined to believe that the latter is decidedly the more correct appellation. In describing the Almond Tumbler, its standard excellencies will be better pointed out and more readily understood, if expressed in the same systematic manner as was adopted in regard to the Pouter and Carrier.

a. Size and Shape.—This property appears naturally to stand first in order, although many regard the succeeding one as of more importance in the Almond. In size it should certainly be one of the smallest of its kind—too diminutive it can scarcely be, according to the prevailing fancy: the body should be round and compact; back very short; neck slender but not long, and carried well curved inwards, exposing prominently a fine ample breast: the shorter the bird stands upon the legs, the neater and smaller the feet or toes, and the shorter the wing-flights and tail feathers are, the more the bird is esteemed. It is more particularly in the points comprised in this property that the *breeding* of an Almond may be detected.

b. Head.—This should be as broad as it is long, and as high as it is either,—in point of fact, it ought to be as nearly as possible round like a ball: an elongated head tapering thin in front towards the bill, is despised as being " mousey " shaped. The front of the head (or forehead) should project or hang over the commencement of the bill in a way hardly describable, but which when accompanied by a short bill and full puffy " chaps " (or cheeks) on either side, gives the bird what is termed a " short-face,"—a property much admired and ardently sought after and cultivated by the Fancy.

c. Bill.—In shape resembling that of the Gold-
finch, it should still be very straight, fine,
and pointed; it cannot well be too short,—if
possible, not more than half an inch measured
from the outer rim of the iris of the eye to
the point, or as we should rather describe it,
three-eighths of an inch in the gape, although
the former is the mode of measuring adopted
by Fanciers: when not more than five-eighths
of an inch long from the eye, it would be allowed to be first-rate, but if it exceeds three-
quarters of an inch, no matter how good the
other points were, the specimen would be
rejected.   The "wattle," or skin across the
base of the upper bill, must be thin and small
in quantity, and resembling a stout thread
drawn over, quite clear, and not covered or
overgrown by the short feathers of the forehead: this little point, even, adds much to
the beauty of the bird in the estimation of the
true Fancier; giving what is termed a good
"stop" to the bird's forehead.

d. Eye.—The eye should be very full, prominent, and bright, the iris being of silvery or
pearl-like whiteness; it must not be surrounded by any margin of naked skin, but
the outer rim should be thin, clear, and
round, being encircled quite up to the edge
by the short close feathers of the head.   The
best position for the eye is exactly in the

centre of the round head; but if seated a trifling degree *below* and very slightly more *backward* than the central point of the head, it is more admired.

e. Feather.—We now come to a property which, though an important consideration with the Fancier, is too often sought to be obtained to the neglect or absolute sacrifice of the preceding and principal points: not that we in any way undervalue the increased beauty of a bird, which in addition to them, possesses also the beautiful feathering which, as an *adjunct* of beauty, it should certainly be the aim of all amateurs to attain in the highest perfection. The ground or body colour should be of a rich yellowish buff hue, deep, bright, and clear, somewhat similar in shade to the plumage of the Golden Sebright Bantam : some have described the colour to be of a " rich Almond " shade, and have hence accounted for the application of the popular designation of " Almond " Tumbler. The most knowing members of the Fancy contend that the ground-colour ought to be a " rich bright yellow;" but this it is conceded is scarcely (if ever) attainable, properly intermixed with the other colours required; and it therefore seems to us absurd and useless any longer to retain a standard which it is not practicable to reach in breeding. Difficult as it is to obtain a pure ground-colour, it is still more so to procure a bird possessing this plumage delicately but distinctly pencilled or broken with a deep black and clear white: this peculiar marking it is indispensable should be present (to constitute a good specimen) on each of the nine first flight feathers of the wings, and also in the tail; but a *perfect* Almond, according to the highest standard, would also have the back, breast, and under-parts variegated in a similar way; whilst the hackle feathers of the neck should have the black ermine-like patches or spangles especially well developed, and of a glossy lustrous shade,—indeed there are some Fanciers who assert that they have succeeded in producing birds with nearly every feather of their plumage broken with the three standard colours.

It may be well to remark, that however good the bird may be, it does not generally attain to the full brilliancy or beauty of feather until the third moult,—up to which period also the colours of the plumage sometimes change considerably: occasionally too they will run from good mixed colours into indifferent " splashed," mottled, and even whole-coloured birds. So entirely do the Almond Tumblers appear to be the produce or result of the highest and most delicate and careful breeding, that even first-rate specimens cannot be depended upon to produce correctly feathered progeny. The Hen is of somewhat smaller and more delicate proportions than the male; but in plumage she is much inferior, the colours being less bright, and the beautiful markings (so highly prized) not so well defined.

25. *Beard.*—These should possess all the points of Tumblers, differing only in the plumage, which is usually of one entire colour excepting the flight feathers of the wing, the tail, and rump, which must be of a clear white. They are distinguished by a small regular patch of white feathers (which we may not inaptly call an " imperial ") extending below the lower bill, and upon the throat, forming a pretty contrast with the rest of the plumage: this *beard*, however, must not be confounded with the ruff-like appendage, so called, seen upon the Poland fowl,—that of the Beard Tumbler Pigeon does not consist of any external appendage of feathers, but is simply formed by the *colour* of the throat feathers being different to the body, which are of various whole colours, as yellowish-buff, red or cinnamon (light and dark), slaty blue, and black. At the last Metropolitan Exhibition, Mr. A. Ball, of Nazing, Essex, entered some " Silver " Beards, which we presume were of white or grey plumage,—if so we fancy that the chief beauty of the Beard must be lost as regards feather. There are a great many coarse

long-faced Beards often seen, with bills running an inch and a half (nearly) from the outer edge of the eye; these are capital flyers, but find no favour with Fanciers: good specimens of short-faced Beards are scarce.

26. *Bald-pate.*—Like the preceding the general ground plumage of this Tumbler is invariably of one whole colour, with pure white flights, tail, and hinder parts. It receives its name from the feathers of the head and throat being entirely white, thus giving the bird the appearance of a *bald head.* The light red or cinnamon with darker shaded necks are generally most esteemed as show birds, but the blues are reputed to be the best and highest flyers: all, however, are uncommon and prized. They are hardy and strong in flight.

27. *Helmet.*—This sub-variety (now very rarely met with) answers nearly to the description of the Baldpate Tumbler, except that the arrangement of the colours of the plumage is *completely reversed:* thus the ground colour of the Helmet is *white,* and the flight feathers, tail, and rump, of some *whole colour,* as yellow, red, &c.: the feathers on the head, also, are of a corresponding colour to the flights, tail, &c.,—the head appearing as if it were covered with a coloured cap or *helmet,* —and hence its name. Our readers will at once observe that the Helmet has *coloured* plumage precisely on those parts which are *white* in the Baldpate, and *vice versâ.*

28. *Smiter.*—This sub-variety, if it ever existed as distinct, has now entirely disappeared; but we strongly suspect that it never soared beyond being a coarse, large, ill-bred, mongrel Dutch Tumbler, mixed perhaps with the Runt family. The only account we have of it is in the *Pigeon Fancier*, by D. Girtin, 1802,—in which it is described as being "in shape, make, and diversity of plumage, nearly like the Tumbler,—size excepted, it being a much larger bird: when it flies it has a peculiar and tremulous motion of the wings, and commonly rises in a circular manner . . . . . and though it does not tumble, it has a particular way of falling and flapping its wings." It is from this habit of *smiting* its wings together that the vulgar appellation of " Smiter " has been given to it. It is supposed to be the same pigeon as the Dutch call *Draïers* or *Dragers*.

29. *Skinnum.*—This is unquestionably a cross-bred bird, descended from a Tumbler and a Dragoon or bad Horseman—its sharp flattened head, long bill, and wattled eye, at once indicating the presence of Carrier blood. It is perhaps owing to the loose *skinny* eyes and bill (so totally opposite to the Tumbler points) that they are vulgarly called " Skinnums."

III. *Columbæ Cristatæ.*

CRESTED, or TUFTED.—It is singular that the

pigeons which are included in this and the succeeding group, and which we have now for the
first time formed into a separate class, have never
hitherto found much favour either with the naturalist or the fancier,—having been passed over
altogether by the former, or dismissed with a very
brief and cursory notice, whilst by the latter they
have been regarded as scarcely worthy of the care
and attention ordinarily required: although assuredly no other varieties seem to us to possess
better claim or title to be looked upon as a primitive and distinct race, than those exhibiting such
a striking peculiarity of plumage, as the addition
of a tuft or crest of feathers projecting outwards
beyond the others at the back of the head, and
sometimes, as in the next group, accompanied by
a ruff-like extension of those projecting feathers,
running down beneath the throat.   The consideration of this subject would afford much interesting
matter for inquiries which might be pursued further, perhaps not altogether without advantage, if
our limits did not prescribe the range within
which our observations must be confined.

30. *Trumpeter.*—This variety (which is at present rather uncommon) is of very Runt-like form
and size. Plumage generally of a pure white,
but sometimes black mottled. It is described,
like other crested varieties, as being distinguished
by a " turn-crown," or tuft of feathers at the back
of the head ; but this is not all, another tuft springs

from the forehead, just above the base of the bill, and the legs are heavily feathered down to the toes. Of its singular name various explanations have been attempted : some say that it is derived from the peculiar trumpet-like sound of the cooing, which we confess we never could detect,—some that it is from their "military air" and "soldierly appearance,"—whilst others ascribe it to their being originally introduced into Holland by a drummer or *trumpeter*,—all more fanciful than probable. May not the curious designation be an allusion to the altogether *unique* appearance of the feathery tuft projecting, like a *horn* or *trunk*, from the forehead, as it would be described by our continental neighbours? thus *trompe-à-tête* is not a very unfitting term for it, denoting a head with a *trunk* or proboscis, and the sound of it bears a striking similarity to the English cognomen of *Trumpeter*. This frontal tuft, the writer of the old *Treatise* already mentioned very properly observes, " is the proper characteristic to know the bird by; and the larger this tuft is the more they are esteemed."

31. *Finikin.*—This pigeon, saith the *Treatise*, is " in size, shape, and make very like a common Runt ;" and we can only further add the assurance the writer gives us, that it " has a tuft of feathers on the hinder part of the crown, and is always black or blue pied." We are further informed, upon the same authority, that its peculiar charac-

teristic is to rise a little way from the ground, when courting, turn round three or four times, flap its wings, and then turn as many times the reverse way! This supposed remarkable characteristic of the Finikin we believe to be common to almost all *flying* pigeons when confined—at least we have observed it in our own loft with many other pigeons,—and it is nothing more than a muscular extension of the pinions as a substitute for flying exercise. The *Turner* is also mentioned, and is described as " in many respects like the Finikin, except that when it plays it turns *only one way*,"— but for this important difference, it seems they are (like the niggers Cæsar and Pompey) so much alike, " especially Pompey," that they can scarcely be distinguished. However, if they ever existed, there are certainly none such known now.

32. *Barbary,* or *Barbe.*—This elegant little pigeon is supposed to have come originally from the country after which it is named, in northern Africa; but it has probably been domesticated with us for a very long period, as it is mentioned by Shakspeare, who makes one of his characters exclaim that he will be " as jealous as a *Barbary* Pigeon!" In size it does not exceed a Tumbler— but the smaller the better; and it may be said to be of a very pretty and smart appearance. The head is of a wide squarish form, not much unlike that of the parrot; this, in our opinion, when true-bred, is always surmounted with a small prettily

curled "turn-crown" of feathers at the back,—though there have been some very fine birds shown that did not possess that tuft: the bill is short and very stout, with only very little development of the fungous skin at the base: the eyes are preferred with irides of pearl-like white, surrounded by a rosette of naked fungous skin, of a granulated surface, and a bright deep-red colour; this, indeed, forms the chief distinction of the Barbe, and when of a large size, it adds much to the beauty of the bird by the pleasing contrast to the plumage: these rosettes are seldom large in young birds, nor until the third or fourth year do they attain the full size, which, however, rarely equals the size of that around the eye of the English Carrier. The colour of the plumage is most esteemed when of a dun or a black shade; they are also found of white, yellow, or red, and sometimes pied or mottled, though these latter are little valued, and thought to be produced from a cross with the Mawmet. Barbes possess a strong and rapid flight; but they are delicate to rear, and thence may have arisen their scarcity, for although formerly much admired, they have now become very uncommon, and good birds are restricted to few possessors: the specimens which have been exhibited by Mr. A. Ball were the finest we have seen, and possessed rosettes of remarkable size and brilliant colour.

33. *Mawmet*, or *Mahomet*.—By some this pigeon

has been thought to be nothing more than a White Barbe, and the writer of the *Treatise* so describes it: but later Fanciers assert that the Mawmet has plumage of a cream colour, barred across the wings with black; the feathers also present the singular appearance of being of a cream shade on the upper surface, and a rusty black hue underneath; the small wattle on the bill, the rosette around the eye, and even the skin, is said to be of a similar sable colour. It is, however, quite unknown to the modern Fancy, if not extinct.

RUFFED.—Under this class we propose to range all those pigeons, which, in addition to the tuft or crest of upturned feathers, are also distinguished by a frill of ruffled feathers, extending more or less down and around the neck, or merely beneath the throat, descending in a ruff-like patch upon the breast,—marks which appear strongly to separate them from other pigeons in which they are absent.

34. *Turbit*, or *Owl.*—Although usually designated under these different names, we have not hesitated to place them together, believing them to be synonymous: certainly we never could discover any distinctive difference between them, nor have we ever met a Fancier who could point out any,—indeed it is admitted that the two " are nearly allied and closely resemble each other;" whilst our continental neighbours class them together as *Pigeons Cravates*, in allusion to the frill

on the neck. The English name of *Turbit* has been attempted to be explained as a corruption from the Dutch cognomen of *Cort-beek* (Short-beak), though we confess we cannot trace it: a far more likely derivation of the name would seem to be the Latin participle *turbatus*, which properly describes the *disordered* and *ruffled* appearance of the feathers upon the breast. In size it does not exceed the average of a Tumbler, but has comparatively a much larger and squarer shaped head, rounded on the top; it differs from the *C. Cristatæ*, being destitute of a crest, and including the Turbits among them, may be thought a very Irish act,—but in the absence of any scientific disposition of the various groups, it would, perhaps, be difficult to find any more appropriate class in which to place them. The bill of the Turbit should be as short and stout as possible, " like that of a partridge ;" and herein some have sought to constitute a difference in the Owl Pigeon, which is said to be alone hooked on the upper part, like the bird after which it is named ; but the fact is, we have seen ill-bred Turbits having very hooked bills, and so-called Owls with good straight bills. The distinguishing characteristic of this breed is the " purle " (*fringe* or *embroidery*) of feathers upon the breast, which has a peculiar ruffled or frill-like appearance, from the feathers " opening and reflecting both ways," curling upwards and outwards laterally—the more full and ample the better; from the lower bill a "gullet" of twisted feathers should descend upon the throat and meet the "purle." In respect of plumage, the Turbit is exceedingly varied; some are whole-coloured white, red, blue, and dun; others (and these are by far the prettiest) may be termed pied, having white feathers all over the body, except upon the upper part of the wings (the flights being white) and tail, which are either of a blue, dun, or black shade; sometimes yellow or red, when the tails should be white like the rest of the body; the blue pied have black bars across the wings in addition. Among the Fancy the Turbits are distinguished according to the colours, as yellow-shouldered, red or brown-shouldered, &c.

35. *Nun.*—This is also a diminutive breed: the proportions of the Nun are Tumbler-like in size, though not exactly in shape; it is an exceedingly pretty and showy bird in the cote, and very fair flyer withal. The plumage is entirely white, except upon the head, the flight feathers of the wings, and the tail, which should always be of a whole uniform colour, as either all yellowish-fawn, red, blue, or black. The coloured plumage which covers the head of the Nun gives it the name, from its resemblance to the "veil" or "hood" worn by those religious ladies: the feathers of this hood are slightly reversed or curled upwards at the ends, around the lower part and back of the head, forming a small tuft; and from the feathers

of this tuft at the back being white, tipped *under-neath* and meeting with the snowy feathers of the back of the neck, (which are also a little ruffled,) many writers have been led to describe the Nun as having a " white tuft ; " but we are satisfied that good clean-feathered birds should have the hood and tuft of a *uniform whole colour*. According to the shade of the coloured plumage, they are styled yellow-headed, (query *hooded ?*) red-headed, &c. Temminck speaks of some with plumage *the reverse* of what we have described, the body being *black*, with the head and flights *pure white :* surely he must have mistaken a Bald-pate or Helmet Tumbler for a Nun ; at any rate no such Nuns are found in this country. The head and bill should be small and neatly formed, and the eye bright with a pearl-white iris.

36. *Ruff*.—This is often described as a *bastard* strain, though we think it may with more propriety be looked upon as the original type of the breed now known to Fanciers as the Jacobine. The *Treatise* states, that it " is in feather, shape, and make much the same as a Jacobine, insomuch that they have been frequently sold for such : " and at the present day they seem to be entirely absorbed in the breed for which it is thus said to have been passed off, for they never make their appearance at the exhibitions. The similarity between the Ruff and the Jacobine obviates the necessity for describing it, as it possesses all the features of the

Jacobine, though in an inferior and less perfect degree, and united in a body of coarser mould and larger proportions.

37. *Jack*, or *Jacobine*.—This pretty little pigeon is so named from the ecclesiastical hood of feathers which surrounds the head, forming its principal characteristic, and resembling the cowl worn by the Jacobine monks of the Dominican order. In size it should be the smallest of any—not even excepting the diminutive Tumbler. The head should be small, the bill short, and the irides of the eyes pearly-white. The distinctive hood consists of a compact mass of ruff-like feathers, encircling the back of the head, and running down in a continuous " chain " on either side of the neck till the ends almost meet under the throat, whence it returns, somewhat after the fashion of the lappets of the lord chancellor's wig, forming a smaller ruff upon the top of the wing-shoulders : this chain (which is called the *cravat* by the French) should be luxuriant, but at the same time thickly and closely curled, and lying close to the head,— a loose flowing sort of chain (or *mane* rather) being a great defect. For the matter of the plumage, good specimens are of every colour— yellow, red, brown, black, and mottled or pied ; but the yellows are more admired, simply, perhaps, because they are scarcer : of whatever colour, the head itself, (not the ruff,) the wing-flights, and the tail, should be of a clear unmixed

white. To our own fancy no kind is so rich or hand-some in plumage as the pure lustrous black, with fine metallic shades, contrasting with the snowy-white of the head, &c. Jacobines are also known under a variety of trivial names, as Cappers, Ruffled Jacks, &c.

38. *Capuchin*, or *Carmelite*.—This is larger and in every respect inferior to the last named: it is said to possess a tolerable hood upon the head, but is destitute of any chain upon the neck. It is probably only a cross-bred bird, as the writer of the *Treatise* assures us, that " a Jacobine and another pigeon will breed a bird so like a Capuchin," as to render it extremely difficult to distinguish.

With this our account of the *Columbæ* closes; having, it is hoped, not omitted any individual member worthy of notice, or known to the English Fancy. It is true that occasionally we see named pigeons with very grand and high-sounding titles,—such as Hyacinths, Scagliolas, Blue Brunswickers, Red-breasted, Dresdens, Pisa, (query Italian Runts?) &c. &c., although we confess to a total want of acquaintance with these: still we are disposed to think that they will probably be found referable to one or other of the varieties enumerated above, with possibly some additional peculiarity, such as an accidental variation in the plumage, or otherwise; in which case we trust that our descriptions of known and established kinds are sufficiently clear and ample to enable the amateur at once, and without difficulty, to assign to them their proper place in the list.

Among French naturalists the following kinds are mentioned; but they are either so carelessly or inaccurately described, as that they cannot be recognised, or their existence is apocryphal.

Norwegian,—a large, white, feathered-legged bird.

Swallow,—distinguished by its " plunging or sailing in the air, when flying."

Tambour or *Glou-Glou*,—from its peculiar note, (perhaps Laughing Runt,) feathered-legged, and all colours.

Jacinth, (query Hyacinth ?)—slaty-blue, and pied on back and wings with white.

Batavian warted, (wattled ?)—short body, long legs and neck; " tall enough to drink out of a pail without trouble."   *Crede !*

Looking-Glass,—of two colours only; " blood-red or yellow wings barred with white."

Mountebank, or *Pantomime*,—nothing more than Gallic names for Tumblers, which they say were bought up by the English in 1817, to improve the English breeds.

Swiss and *Polish*,—kinds of Tumblers; the latter said to be " crested."

Ord. NATATORES; Gen. ANATIDÆ.

OF this numerous tribe of birds, which comprises an almost endless variety, we purpose only to describe such as have become domesticated amongst us, and may be fairly included in the same general term of Poultry: of the wild species, indeed, we possess but meagre knowledge; our acquaintance with them extending little beyond the occasional visits which these migratory wildfowl pay us, when, driven from their usual haunts by the severity of a northern and eastern winter, they seek a refuge in our scarcely more genial clime.

SWANS.

Sub-genus Cygnus.

WILD SPECIES.—Although the Swan has ever been a very scarce bird with us, it abounds upon the borders of the great eastern inland seas, and inhabits the secluded lakes or undisturbed swamps. Northward they are also found in localities of a similar nature as far as Lapland; whence they come in flocks (though not in great numbers) to our shores, upon the setting in of the intense frosts and cold of winter. The wild species with which naturalists are best acquainted, is the

Hooper or Whistling Swan, (*Plate* VIII. *fig.* 1,) so called, it is supposed, from the singular power it possesses of uttering a sort of melodious *whoop* (whence its name) or guttural whistle, owing to the peculiar organic structure and recurvation of the windpipe. This Swan is the largest of the European wild species,—measuring from the point of the bill to the tip of the tail about 5¼ feet, and across the wings (expanded) upwards of 7 feet; weight from 23 to 25 lbs. The plumage of adult specimens is invariably pure white, though in young birds there is a considerable admixture of dusky ashy-coloured feathers. The bird is distinguished by the bill, from the base rather more than half way down to the point, being of a yellow colour, whilst the remaining part to the point is black; the legs, toes, and webs (or membranes) are entirely black. — *Bewick's Swan,* so called in memory of the artist whose master-hand so beautifully illustrated the natural history of birds, was first introduced as a variety distinct from the Hooper, by Mr. Yarrell, about the year 1828, when a splendid specimen shot at Great Yarmouth, in that severe winter, came into his possession. In size it is less than the preceding, not exceeding 3 feet in length; in plumage and externally it differs little from the Hooper; but the anatomical structure, especially of the organ of voice, forms the distinction by which scientific men have separated it. Upon the water it is less graceful than either the wild or domestic Swan, and more resembles the Goose,—indeed it has been named the *Swan-Goose* by some, who have been of the opinion that it holds a sort of intermediate place between those two birds.—The *Polish Swan*, though for a long period known to the London dealers, (who obtained some from the Baltic,) has only very recently attracted the notice of naturalists, and we are indebted to Mr. Yarrell for our acquaintance with it. It is but little inferior in size to the Hooper, but in form, habits, plumage, &c. it seems closely allied to our domestic variety, the organ of utterance being similar.—The continents of Australia and America furnish other and distinct varieties: the former possessing the black-necked Swan, and the black Swan, whilst the latter has the North American Swan, and, the largest of all, the Hunter's Swan. We believe there is no authenticated account of any of the wild Swans breeding with our domestic species in this country.

DOMESTIC SPECIES.—Our half domesticated *Mute* or silent *Swan*, (*Plate* VIII. *fig.* 2,) is at once the largest and most graceful of all British birds: and the antiquity of this majestic and stately bird is exhibited in the pages of history and of poetry. Although it is said to be found in a wild state in many eastern European countries, it is probable that the original has been lost in the lapse of time. In this country they are chiefly found upon the

Thames, the river Yare, and probably, as in former times, upon the inlet of the sea near Abbotsbury, Dorset, and in the Trent. In form the Mute is scarcely so elongated as the Hooper, although the body is perhaps larger and more compact, the whole weight of a living specimen ranging from 25 to 28 lbs: the plumage also is white; and the principal external distinction from the wild is in the colours of the bill, (which is also somewhat larger,) for whereas in the wild the basal part or half is yellow, and the remaining portion to the point is black—in the Mute Swan they are exactly reversed, the basal half being black, and the other orange-red, with a prominently projecting tuberculous knob, (or " berry," as it is termed technically,) which is borne in front on the forehead, and is of a black shade. Internally it likewise differs materially from the Hooper, whose capacity of voice it does not possess; though it is not the absolutely *silent* bird which its name would import. Every one is aware of the poetic notion that has always prevailed, of the Swan uttering a melodious but melancholy cry or lament just before its death: an idea thus simply expressed in one of our Old English madrigals;

" The silver Swan, which, living, had no note,
    When death approach'd, unlock'd her silent throat : "

and assuredly the Mute Swan does possess the

power of uttering a low and plaintive wailing note, which Mr. Yarrell positively states he has heard at various times,—though it is quite possible the note may be more distinct and continuous as a sort of departing dirge.

GEESE.

Sub-genus Anser.

WILD SPECIES.—In our country the Wild Goose is a bird of passage only, and the harbinger of cold weather. At the close of the year these migratory strangers may be seen in flocks, more or less numerous, shaping their course from more northern climes to the various localities on our own coasts, congenial to their habits, or capable of supplying their alimentary wants : their early appearance here being taken as a sure indication of the intensity of the cold in their native habitats, and of winter's approach hitherward. There are few persons but have witnessed the (to our minds) extremely beautiful sight afforded by a flight of these hibernal visitants, winging their way at a considerable altitude overhead through a clear and frosty November sky, their presence being made known by the clarion-like clang of their not un-musical cries mingling in shrill unison. The order they observe in their migratory flights is peculiar : if the flock comprises any considerable number of birds, they range themselves into two lines, each projecting a little in front, and laterally of the other, with general leaders to the two divisions, forming an angle of wedge-like shape, thus, ➤, which they preserve with great regularity ; if, however, the number is small, then there is only one single line, called a " string," the order of which, however, is maintained with equal regularity. Under the general and popular term of " Wild Goose," are included several distinct varieties,—to attempt a description of these, or to seek to trace from them the origin or descent of our domestic race, would be entirely beyond our design ; and certainly we have not even a remote intention of entering upon such a " chase " as that to which these birds have proverbially given rise, and which the subject would involve. We shall therefore content ourselves by simply enumerating the various divisions in which the wild *anserine* tribe is believed to be comprised ; their history may be sought in the works of Yarrell and other eminent ornithologists.

GREY LAG, or Grey-legged, Grey Goose, Feral or Wild Goose.

BEAN, or common Wild Goose.

PINK-FOOTED, or Short-beaked Goose, Mountain Goose.

WHITE-FRONTED, or Laughing Goose.

BERNICLE, or BARNACLE, Gannet, or Solan Goose.

BRENT GOOSE, or Ware Goose.

RED-BREASTED.

EGYPTIAN, Fox Goose, Chenalope, or Cairo Goose.
SPUR-WINGED, or Gambo Goose.
CANADA, Cravat Goose, or Swan Goose.

Of the *Grey Lag* (generally considered as the common stock whence our domestic race is derived, but whether rightly or otherwise we take not on ourselves to determine) it may be necessary to give some brief notice. To Mr. Yarrell we are indebted for a restoration of the old name of Grey *Lag*, (supposed to be derived from the Italian *lago*, or the English word *lake*, as denoting their favourite haunts,) a far preferable name to the more popular one of Grey-*legged*, which it certainly cannot be said to be. It was formerly very commonly met with in the marshes and swamps of the Eastern and Fen counties, and also in Ireland; but the progress of drainage and cultivation has so gradually absorbed their hibernal resting-places, that they have now become exceedingly rare visitants; they are more common in Germany and Central Europe, and likewise occur in Northern Asia; they go as far north as Iceland in summer, but have not been observed on the northern part of the American continent, notwithstanding their wide geographical range. In form the Grey Lag Goose is perhaps rather more elongated than the common domestic Goose, and rather exceeds it in size: the general plumage is of a dark ashy-grey shade; the feathers on the head, neck, and back being of a greyish brown, marginated generally with lighter grey edges; those of the wings and coverts shaded with light and dark leaden-grey; tail, tail-coverts, and rump, very light grey or white; belly and under and hinder parts white; bill, legs, and feet of a dull pinky-flesh colour.

The other varieties are distinguished either by some external differences, (indicated in some by their name,) or by some peculiar organic structure: but in addition to these, other naturalists mention nearly twenty supposed distinct " varieties," inhabiting various parts of the Northern and Arctic regions, and of the continents of India, Africa, and America, &c.; but it may be reasonably suspected that very many of them are far too nearly allied to those we have already enumerated, to be entitled to claim at our hands any separate distinction or classification.

DOMESTIC SPECIES.—As we have intimated, the origin of this useful tribe of domestic fowl is by no means determined by naturalists; for whilst the majority refer to the Grey Lag as the primitive stock,—the possession of points of resemblance in the domestic to other wild species, induces some to assert that we are indebted for its production to more than one wild variety. Be this as it may, it only remains for us to describe the Common Domestic and its varieties, or, more correctly speaking, sub-varieties.

Common Domestic.—The Geese which have

hitherto passed under this very general designation, do not appear to have received much attention in their breeding; and as a necessary consequence, those possessing any distinct shade or variation of plumage originally, have become so intermingled, that they can scarcely now be separated, amid the diversity and blending of feather produced. There does not, however, seem to us any good grounds for dividing this class into the numerous sub-varieties it is usually ranged in: and in the absence of any specific or more marked distinction, beyond what is at present ascertained, we believe all its members may be comprised within the following limits:—

White,—Common, and Embden or Irish.
Grey,—Common, and Toulouse or Mediterranean.
Mottled,—Common Grey and White, and Saddle-backs.

The *White* appear to be the largest, if not the heaviest, of its kind,—and some may even add, and the *prettiest* too, though that must clearly be judged as a matter of taste only; the entire plumage should be clear and unbroken white. The Embden takes its name from the Hanoverian town whence it was, many years since, imported; and whence, as also from some parts of Prussia and Holland, we still continue to draw supplies. It differs in no respect from the common English Goose, having precisely the same shape and form, the same pure white plumage, the same rich red bill, legs, feet, and webs,—indeed, although it has been dignified by the title of a distinct variety, it modestly puts forth no such pretensions itself—the honour has clearly been thrust upon it. It is certainly bred in a superior manner to our own kinds, and attains a greater size, the net weight of the gander ranging from 16 lbs. to 19 lbs., and the goose 12 lbs. to 15 lbs.; this additional superiority of size, however, in most specimens we have seen, has been more than countervailed by the extra offal, arising from the increased length of neck and legs, though, possibly, this may not generally be an objection. The fine specimens exhibited some time since by Mr. Nolan, at the Dublin Show, were nothing more than remarkably good white birds, notwithstanding that they were styled *par excellence* "Irish Geese," by their owner,—a name only calculated to add to the difficulty of preserving *nominal* identity of the pure white breed. (*Plate* VIII. *fig.* 3.)

The *Grey*, like the preceding, really and truly constitute but one class, the recent importation of the Toulouse consisting only of birds identical in kind and character with our own common Grey Goose, though undoubtedly bred in a very superior way, possessing also points of surpassing excellence, and attaining to a most extraordinary size: their introduction cannot fail of very greatly improving our own strain. In describing the Grey Goose, we cannot do better than take for our model

one of the favourite *Toulouse*, which we regard as
the best specimen of a grey that can be produced.
It was imported some few years since by the Earl
of Derby, from the shores of the Mediterranean,
and, it is said, from Marseilles, where, as also in
other parts of France, the breed had long been
celebrated among good poultry-wives; how it
received its foreign name is not quite apparent.
Size alone does not distinguish this sub-variety;
its round compact form, capacity for stowing fat
away on its frame, gives it just the right form for
such poultry; the head is large, long, and thick,
and slopes down from the crown perfectly straight
to the bill, there not being (as Pigeon Fanciers
would say) any " stop " between the forehead and
bill; the neck shorter and thicker than the common; the depth of body from the back to the
breast and belly is very great, and at the latter
point this appearance is increased by the abdominal pouch, which in the Toulouse is very early
and conspicuously developed, in adult specimens
it nearly sags down upon the ground; in short,
in form the Toulouse is the Cochin China among
Geese, short, round, and compact, and yet of the
largest proportions withal: the plumage should
present great uniformity of colour in the disposition and marking, and generally the different specimens do; the colour of the feathers on the head
and neck, a bright, dusky, ashy-grey—the front
part being of a lighter shade, as also is the breast;

the back, wings, and flanks of a rich dark olive or greyish-brown, exhibiting various degrees of shading, and edged or slightly mottled with white, the shafts of the feathers being likewise white; the tail coverts, under and hinder parts, pure white; the tail is short, consisting of brown feathers banded with white at the ends; the legs, (which are short and thick,) feet, and webs, of a deep orange-red colour, with dusky or olive claws or nails. Amongst our English breed of the Grey Goose, numerous specimens are to be found which answer very nearly to the above description, especially as regards the colours and markings. The introduction of this excellent foreign breed of Geese cannot fail greatly to improve our own, but the supposed *distinctiveness* will not be long ere it disappears, when the correctness of our assertion will be demonstrated,—namely, that the Toulouse is only a superior strain of our own common Grey Goose.

Mottled.—We have had many doubts as to the propriety of distinguishing these, even as a subvariety only, from the heterogeneous stock of our country poultry-wives; but as we are satisfied that they are the production of the White (or Embden) with the Grey (or Toulouse), and have now very generally established themselves with us,—it may be well, perhaps, to concede them the honour of a passing notice. In size they are of fair proportions, and differ but very slightly from those previously described, save in the plumage, in which white predominates, mottled in patches with a greyish-brown, only irregularly. The geese of this kind most valued, as well for their appearance as utility, are what have long been known and esteemed as " Saddle-backs," so called from a large broad patch of the dark feathering upon the back and wings. They can scarcely need further description.

Chinese Goose.—This variety (for such there can be no doubt it is) is known under very many different names; thus we have the following, all indicating the same bird:—

African Swan,	Muscovy Goose,
Asiatic Goose,	Northern Goose,
Chinese Swan,	Polish Goose,
Cygnoides,	Russian Goose,
Guinea Goose,	Siberian Goose,
Hong Kong Goose,	Spanish Goose,
Knob Goose, &c., &c.	Swan-Goose.

This prolific nomenclature affords an illustration of the facile way in which innumerable " varieties " of the Goose tribe are created out of one unfortunate and unoffending family, and thence find their way into the Poultry Books as such. As the endless designations bestowed on them indicates, the habitat of the Chinese has not yet been satisfactorily ascertained : it is certainly equally a native of Africa, China, various parts of the Asiatic continent, and has been found, it is said, in New Zealand; whilst undoubtedly we have frequently been supplied with them from the Mediterranean countries. This variety was known to our old writers on such matters, and is now sufficiently plentiful in Great Britain to be considered one of our domesticated acquisitions, and indeed, according to report, it is said to unite well with our own common sort. Formerly much difference of opinion existed among naturalists as to whether it really was a Goose—many concluding it belonged to the Swan tribe; but there can be no doubt, on a further acquaintance with its structure and habits, that it belongs to the *anserine* family; although many assert that, like Bewick's Swan, it holds a middle position between the Swan and the Goose. Some writers have constituted into separate classes or sub-varieties certain of these birds, having only very trifling differences in the plumage, &c., and which is often, we think, dependent on age, sex, or accident: and at the risk of differing from most who have preceded us, we shall select the *White* Chinese Goose as the type,—believing that it is the most distinct in its character, as it certainly is the most handsome in appearance, of its kind, and, moreover, it presents a better title to the prefix of " Swan," which enters into the composition of so many of its names. In this view, therefore, we should say, that the prevailing colour of the plumage is a clear, pure white; the distinctive mark in this, as well as in the darker sub-varieties, being a longitudinal streak or band of a dusky brown shade running down from the head, along the back of the long neck, almost unto the back: besides which it has a characteristic berry or "knob" of a tough fleshy excrescence seated at the base of the bill, somewhat like the Swan; and under the throat the skin forms a very singular double-folded flap or pocket: the bill, berry, and legs, *in the pure white*, should be of a bright deep orange-yellow hue, notwithstanding what has been said about the colour varying in those parts. All the members of this variety are remarkable for their erect carriage upon land, and graceful appearance on the water; and no less for the harsh, disagreeable, and incessant screaming clang, which they utter with annoying perseverance, during the entire day, and often at intervals in the night, without any apparent incitement. The male (especially of the white) is strikingly and disproportionately larger than the female—fully one-third more.

Sub-varieties.—These are usually divided into three, but we shall content ourselves by comprehending them under but one, which may be appropriately styled the *Grey Chinese Goose*, (more generally called *Brown*,) and which are much more common than the pure white. The prevailing colour of the plumage is a grey, relieved by the characteristic stripe of dull brown at the back of the neck; the back, wings, and tail being also of a similar colour, though the shade of the latter is usually lighter; the front of the neck, breast, and flanks cannot be better described than as of a "fawn" colour, whilst the under and hinder parts are of a greyish white: the bill, berry, and legs are mostly found of a dark sooty-black hue,—though we admit, that some specimens, apparently in other points precisely similar, are found with those parts of a dull red shade; for which they have been forthwith installed as the "*Red-legged*" sub-variety: we regard these simply as the result of the union of the Grey with the White,—the latter contributing the variation of colour alluded to. The so-called "*Hong Kong*" sub-variety appears to be nothing more than some extraordinarily large specimens of the Grey Chinese Goose, the product of either successful or accidental breeding.

DUCKS.

Sub-genus Anas.

THIS, the typical member of the genus *Anatidæ*, might, perhaps, have appropriately preceded the Swan and Goose, than which it may be said to be more within the true range and signification of the term "Domestic Poultry." True it is that

few wild species of fowl comprehend such almost
interminable divisions and sub-divisions as the
Duck, no less than from forty to fifty " varieties "
being named; the mere enumeration of which in
our pages would occupy more space than we have
at our disposal: nor indeed does it seem requisite
to enter upon any description of such, which per-
tain rather to the business of a " Decoy," (now
seldom found except in the counties of Norfolk,
Suffolk, and the fens,) or to the sport of the gun,
—such are amply treated on by Mr. Yarrell and
other competent writers: our duties lie rather with
the humbler branches of the family, with which
by their domestication we have become acquainted.
Many of the wild Ducks differ considerably as regards size and form, also varying greatly in the
plumage; although a modern author has said that
when one has seen a wild fowl, a description of its
feathering will exactly correspond with that of any
other,— an opinion which, though perhaps more
correct in reference to the wild goose, must not be
too generally or implicitly relied on.

Mallard, or *Common Wild Duck*.—This is very
generally believed to be the progenitor of our own
domestic duck; and the ease with which in many
individual instances the young of the wild bird
have been hatched and brought up in confinement,
in a sort of half-domesticated state, gives great
show of probability to the supposed assignable
origin of our domestic duck: although recently a

doubt has been raised, and it has been asserted that our farm-yard ducks are derived from the East Indies, where there is a *domestic* race almost identical with our own; but even if we assent to this latter view, how can we account for the striking differences which occur in the domestic variety? surely, not exclusively by the importation of alleged *domestic* breeds. Those who deny the *wild* origin of the common duck, should not forget that the exceedingly domesticated Muscovite variety has a wild prototype in the tropical regions of South America. The Mallard is widely spread over different parts of Europe, and indeed of the Old and New World. It may be said to be a resident in Great Britain, where from a very remote period it has bred in considerable numbers: in England, however, it is fast disappearing as a permanent resident, being gradually driven from its haunts by the culture of the soil. The extreme length of an adult male Mallard is 25 inches: the prevailing colour of the plumage is grey and various shades of brown; the head and upper part of the neck covered with some very fine brilliant green feathers of velvet-like texture; at this point a narrow white band quite encircles the neck, the lower half of which is of a deep chestnut brown hue; the coverts and wings beautifully marked with rich brown and grey shadings, and terminating at the end of the flights in a very brilliant *speculum* of iridescent purple, barred or bounded by a

streak of black tipped or spangled with white;
under and hinder parts entirely greyish-white;
tail grey, except four middle feathers, which are
of a rich velvety black, and *curled upwards* at the
end in a very peculiar way; bill of a dusky greenish shade; legs, toes, and webs a dull orange yellow; the *claws* of the toes a *dark olive*, forming a
most striking distinctive mark of the *wild* variety.
The female Mallard is considerably smaller in size
than her lord, and is much inferior to him as regards plumage, which in her is of a more sober
character; she is clothed almost entirely in brown
feathers, streaked in the middle and upon the
edges with black; on the wings are similar *spe-
culums*, though scarcely so bright, as those on the
male; the under parts, &c. of a light brown shade;
no tail feathers curled up. There is a singular
fact, though well known to naturalists, worthy of
record, in reference to the plumage of the male
Mallard: every year at the close of the breeding
season in the end of May, he loses the whole of
his splendid attire, and exchanges his brilliant
plumage for a garb exactly like that of the female,
—even the characteristic sexual mark of the curl-
ed-up tail feathers also disappearing: the interest
which attaches to this remarkable change, is further increased by the rapidity with which this new
dress is also laid aside, generally in a few weeks,
(at the commencement of the month of August,)
when the drake again re-assumes his gala dress in

renewed and fresh beauty. A few solitary cases have occurred, in which the Mallard duck has forsaken her dress and disguised herself in that of her consort; but these, it is needless to say, are rare and accidental; whereas the change described in the male birds is a fixed law in the economy of their existence.

DOMESTIC SPECIES.—If the wild family is large, our domesticated species are exceedingly limited in number; and even the few that we possess are almost all comparatively modern introductions among us: thus the common brown duck of our ponds and yards, formerly constituted almost exclusively the stock of this country. The domesticated varieties (properly so called) may be thus briefly noticed:—

ROUEN, or Common.

AYLESBURY.

MUSCOVITE, or Brazilian.

EAST INDIAN (black).

DUTCH.

Rouen, or *Common*.—Ths variety has very generally been regarded as perfectly distinct from our own Common Duck: for our own part, we hesitate not to place them both in the same category at once; for, like the Toulouse Goose, we believe the Rouen Duck is only a superior breed of their kind, and no more distinct from the ordinary farm-yard duck, than is the so-called " Norfolk " Turkey distinct from any other breed of black Turkeys. The Rouen variety is known by several names—as

Rhone, from that department in France; *Rohan*, after the Cardinal of that name; and *Roan*, a word signifying (according to Bailey) " a bay, black, or sorrel colour, intermixed with grey; " the derivation he gives being the French word *rouen*, which not inaptly describes the shades of brown and grey plumage in which ducks of this variety are almost uniformly clothed: and will not this last at once explain the origin of the name of *Rouen*, by which it is more correctly, or at least more generally, designated? Although, from the fact of our receiving the principal supplies of this stock from France, many writers have supposed that its appellation is taken from the town of Rouen; and certainly fowls of this variety were formerly much sought after in that place for our markets, but notwithstanding that they partook of the general superiority for which the poultry produce of the fair wives of Normandy are justly noted, they nevertheless in no way differed from the same variety found in other parts of the continent. The Rouen is of the largest size, the natural weight of an adult drake and duck ranging from 12 to 15 lbs. the pair: the shape is ugly, lumpy, and bulky; the ungainly appearance being increased by the greater development of the abdominal pouch in this than in other varieties. The colour of both male and female respectively very closely resembles the Mallard, but even of a richer tone,—the drake has the same brilliant green head and neck (encircled with a white collar) as the wild one; the lower part of the neck and breast of a similar rich brown hue; the back, shoulders, and flanks, usually somewhat more mixed with grey; coverts and wings almost precisely similar in colour and marking with the same beautiful *speculum* on the latter; tail dark brown, with rather deeper margins of white, and the three or four centre feathers curled upwards; bill greenish (dark); legs, feet, &c., orange-red. The back in plumage even more nearly approaching to the colour of the female Mallard, namely of a quiet sober bay or brown shaded with black: the bill, legs, &c., of rather a darker hue than they are in the drake. Our own common mongrel race of this variety comprise almost every size and shape, and the greatest variety of colouring and admixture of shades: they are only too widely scattered, and pass in different localities under various provincial and trivial names; but they are being rapidly forsaken for the improved Rouen strain.

Aylesbury.—This exceedingly pretty white variety is certainly the most ornamental of its whole tribe: but it has been doubted whether they are to be regarded as distinct from the common domestic duck,—or whether they are only a subvariety originally produced from some chance white specimens, and subsequently, by careful selection and breeding, have acquired a kind of permanent superiority,—a question we will not

venture to determine. It receives its name from
the town in Buckinghamshire, where the breed
has long been celebrated, and is even now most
extensively reared by the cottagers there, and in adjoining localities. In carriage and shape they are
very superior to the Rouen, but rather less in size,
seldom exceeding 12½ lbs. the living adult pair
of male and female. They are distinguished by
their beautiful snowy-white plumage, with some
slight shadings of a lemon hue, which we have
always remarked (more or less) in the plumage of
most *white* birds : the bill should be of a flesh-
like or light salmon tint, but it not unfrequently happens, even in first-rate specimens, that as
the birds advance in age, the bills become stained
with small patches of a dusky olive hue, detracting much from the otherwise pretty contrast the
bill affords to the full white head : the legs, feet,
and webs, should be of a pale orange yellow,—
dark stains being even more objectionable than
upon the bill. The Aylesbury duck possesses
equally beautiful plumage, and is so nearly like
him, that but for her slightly inferior proportions,
and more prominent belly, it would be difficult to
distinguish her from the drake, more especially
until the latter, arriving at full maturity, exhibits
the tail feathers curled up; a mark which more extended observation than has fallen to our lot, may
establish as another distinctive characteristic of
it. Our portraits (*Plate* IX. *fig.* 2) are from prize

birds, belonging to Mr. J. Youell, of Great Yarmouth.

Muscovite, or *Brazilian.*—This variety is now generally known as the " Musk Duck," a name given to it by modern critics (who discard the old appellation of " Muscovy") on account, as they say, of the " *plumage emitting the odour of musk,*" or, as others with equal gravity assert, because of the musky "*flavour of the flesh;*" both these alleged qualities existing on no better foundation than the imagination! The critics were quite correct in assuming that the old name of *Muscovy Duck* was erroneously applied to this variety, *if taken to indicate the source whence we obtained it;* but may not the name have been bestowed upon it for another purpose, for instance, to indicate the *means* whereby, rather than the *source* whence, we acquired it? Our own explanation of its original designation is this: it is most undoubtedly to the merchant adventurers who flourished during the 16th century, when the spirit of enterprise (which tended so much to extend the mercantile greatness of this kingdom) was then at its height, that we are indebted for the acquisition of many of the principal products of the New World and eastern countries; and nothing is more likely than that the great Merchant Companies by whom such novelties were introduced, were desirous of perpetuating their name by prefixing it to that of the article,—thus the " Turkey Company " having imported coffee from the East, it was for a very long period afterwards known as "Turkey Coffee," although every one is aware the berry was not cultivated in that country at all; so, perhaps, the Turkey fowl having been procured in the same way, may have received its name. And so likewise, we think, may have originated the designation of "Muscovy" applied to this variety of duck: in the early part of Queen Elizabeth's reign, a very important branch of the Turkey merchants was styled the "Muscovite Company," and this New-World duck may perhaps have been among their importations; in support of this view we have the fact, that the earliest notice of this peculiar variety occurs in a work of that period, (1570,) wherein it is styled the "Turkish Duck." Later it has been called the Indian, Guinea, Cairo, and Brazilian Duck; and on the continent, at the present day, it is almost universally known as the "Barbary Duck." Its original habitat is the Brazils, where it is found wild, as also in other parts of the tropical regions of South America; there the early travellers described it as "of the bigness of a goose," and correctly noticed its points. In size the domestic Muscovite certainly surpasses all other varieties—even the Rouen: the weight of the drake has been known to reach $10\frac{1}{2}$ lbs., and the duck $6\frac{1}{2}$ lbs.; the pair usually averaging from 13 to 15 lbs. live weight: the body is remarkably elongated in form, and carried in a very

straight or horizontal line: neck unusually thick
and large; head also large and knob-like; bill
neither so long nor so stout as might be expected
from the size of the variety,—but it is much pointed downwards at the tip, and at the base surmounted by a bright red papulous mass or fleshy
excrescence, which, as the drake advances in years,
enlarges and assumes the shape of a round "berry,"
of the size of, and very similar to, a cherry;—from
which an irregular patch of naked and wrinkled
skin, of the same hue, extends behind and around
the eye on either side of the head: the legs have
exceedingly stout shanks, and from having the
singular *appearance* of being destitute of any thighs,
this variety looks as if it stood very low on the
leg.   In describing the plumage, it may be well
to notice that the feathers of the Muscovite duck
do not possess the water-proof or water-repellent
quality, which in the earlier pages of our treatise
we stated was a characteristic feature in the *Anatidæ :* this repellent property, it is said, (and we
are indebted to the Rev. E. S. Dixon for the sug-
gestion,) is the result of some minute structural
peculiarity or other in the texture of the feathers
—as smoothness, elasticity, and rigidity,—and it
is perhaps a kind of instinctive knowledge of their
deficiency in this respect, which gives rise to that
disinclination (almost amounting to antipathy) for
the water which the Muscovite variety mostly
exhibit: the colour of the feathering generally is

a dusky brown, with more or less of black markings upon the upper portion of the body—especially on the back : the head is enveloped, as it were, in a wig or hood of short and slightly twisted or ruffled feathers, which entirely cover the crown, back, and sides of the head, and almost close in front under the lower portion of the bill; these feathers are of a soft velvety texture and a rich dark metallic green hue: the front of the neck a greyish-white, often running into a fine reddish purple or vinous tinge in the lower part : shoulders of the wings white, that colour also usually extending along the flights : the bill, legs, webs, &c. varying (with the greater or lesser depth of the colour of the plumage) from a dusky flesh shade to a dull red colour ; it not unfrequently happens that the bill is banded across with black.—The duck is fully one-third smaller in size, and even shorter on the leg, than the drake : the plumage, too, is always inferior,—though very similar in the disposition and marking : the naked skinny patch, extending from the base of the bill around the eye, is present equally in the female, but she has neither the " berry " rising on the forehead, nor the ruffled hood of feathers on the head : the legs of the duck are much shorter than those of the drake, and having a tolerably large abdominal pouch, this gives it even a more awkward and almost deformed appearance.—Another peculiarity is found in the Muscovite, as, according to some writers, they are *quite mute ;* and although this may not be strictly correct, it is certain that they cannot utter the genuine duck-like "*quack!*" their best attempt at making it having been aptly described as a "hoarse asthmatical sigh" or groan. The pair whose portraits we have delineated (*Plate* IX. *fig.* 1) were selected from prize pens belonging to the Lady Paget, of Sennowe Guist, and were the finest we ever saw.

Sub-varieties.—The following are spoken of, but we have no knowledge of them,—a *pure white* Muscovite, seldom met with, and having a cry as loud as other species of duck,—we should look with suspicion on its parentage ; a *pure black*, rarer and still more handsome than the preceding, the beautiful glossy, velvet-like plumage contrasting prettily with the red tuberculous skin upon the face; and lastly, a *pied* family, comprising some of a brown and white and others of a black and white mixed plumage,—these are not very uncommon, and may be the result of the White and Black Muscovites forming an alliance.

East Indian.—We have already given our readers a list (p. 59) of the numerous *aliases* by which this pretty variety is known, and to these others have since been added. The names of "Labrador" and "Botany Bay" ducks, under which this variety is usually advertised by the dealers, are very objectionable, conveying as they do the erroneous impression that those places are their original ha-

bitats,—whilst there is little room for doubting
that they originally came from the eastern continents of India and Asia; for this reason we have
selected the name of East Indian, by which designation we propose they shall in future be exclusively known,—as, being unappropriated to any
other variety of duck, we shall thereby avoid the
confusion which now exists by the application to
them of names of other varieties: thus the " Labrador" duck, properly so called, Buonaparte assures us is a pied and wild fowl; whilst the true
" Botany Bay" is distinguished by the striking
peculiarity of possessing a bill which ends in a
kind of tough substance of membraneous consistency.   The subject under consideration is very
much smaller than the other varieties of the domestic duck, the weight of a living drake and duck
ranging from 8 lbs. to 9½ lbs. extreme: it does not
possess any feature in form, &c., materially differing from those previously described, and is distinguished by the remarkable sable plumage in
which it is entirely clothed: the feathering should
be bright and lustrous in its shade, with brilliant
green and purple metallic reflections upon the
head, neck, and shoulders; the metallic green
tints prevailing upon the wings and coverts; but
there is the same difficulty in getting specimens of
a pure good *black*, as is experienced in the Cochin
China and some other fowls of a similar shade,—
the black in many instances assuming a rusty

or brassy hue, detracting much from the beauty of the birds; occasionally even this variety has been known to cast partially mottled white plumage in moulting, which we should regard as indicating some impurity in the breed,—at any rate an *unmixed uniform black* plumage should be the required standard in the East Indian variety: the bill, legs, and webs, of a dusky greenish black: tail not very much developed, and the adult drake carrying some few of the middle feathers slightly curled up on the back. The duck is scarcely inferior to the drake, either as regards size or plumage,—herein constituting a remarkable exception to its tribe generally, the female almost invariably showing a marked inferiority to the male in every respect. There is a singular peculiarity this variety alone possesses— that of laying *black* eggs, or rather eggs with shells covered on the upper surface with a film or pigment of sooty-black hue; this, however, only takes place at the commencement of the laying season, for after the first few eggs are laid, the colouring gradually disappears, but they never have so light a colour as those of the common kinds. Our portraits (*Plate* IX. *fig.* 3) were taken from some beautiful specimens, in the possession of Messrs. Youell and Co. of Great Yarmouth.

There are many ducks which assume a sort of nondescript character, and which, whilst they may be considered as domesticated, yet appear isolated from the common and known varieties. There is the *Marsh Duck*, as it is vulgarly called in Norfolk; rather a diminutive kind, very much resembling the Mallard both in appearance and plumage:—they are, we believe, almost exclusively confined to the eastern and fen counties, where they are kept in large numbers in the sedgy swamps of the marshes,—or it might rather be said that they keep themselves, for they seem to exist in a semi-domesticated state, and are wild and unsettled in their habits; the only thing to be said in their favour is, that they scarcely require to have any food, care, or attention bestowed upon them. Upon the coast part of the county of Norfolk, in the cottagers' yards are frequently to be seen specimens of a small *Dutch* breed, of a very pretty and peculiar appearance: the plumage is of a whole colour, either a slaty-blue or dun shade, or else a sandy-yellow or cinnamon, something like the colour of the fur of a tortoise-shell cat: a gentleman of our acquaintance, who had some of this latter colour, called them *Rotterdam Ducks*, having obtained them from that place. The *Call Ducks*, also from Holland, are of very small proportions: they arc only kept or valued, as their name indicates, to serve as *calls* in the " decoys" for wild water-fowl, a purpose for which they are admirably suited, owing to the vociferous and interminable *quacking* which they

sustain with evident delight to themselves: they
are of two sorts, one with plumage of a pure white,
and the other with feathering of a greyish-brown
like that of the Mallard; and the drakes and
ducks of this latter colour are invariably selected
in Norfolk (noted for its wild-fowl decoys) as the
most preferable for the purpose required. Of the
Tufted Tame Ducks, said to have " crests as compact and spherical as any Poland fowl," we have
no experience, and we have only heard of isolated
cases and of rare occurrences: they are probably
freaks of nature, and incapable of being rendered
permanent in breeding.

CHAPTER VII.

ECONOMIC VALUE OF POULTRY FOR THE TABLE—PRODUCE
—RELATIVE QUALITIES OF FOWLS—PROFIT—CHOICE OF
STOCK.

In Britain, where a greater quantity of butcher's
meat is consumed than probably in any other part
of the world, poultry has ever been deemed a luxury, and consequently not reared in such considerable quantities as in France, Egypt, and some
other countries, where it is used more as a necessary article of food, than as a delicacy for the sick,
or a luxury for the table. In France, poultry
forms an important part of the live stock of the
farmer, and it has been said of that country, the
poultry yards supply a much greater quantity of
food to the gentleman, the wealthy tradesman, and
the substantial farmer, than the shambles do; and
it is well known, that in Egypt it has been from
time immemorial a considerable branch of rural
economy, to raise domestic poultry for sale, hatched
in ovens by artificial heat.

It has been a general and popular topic of declamation, that in former and presumed happier times, our small farmers' wives raised a superior quantity of poultry to that which has been produced of late years; a position, at best, very questionable, since poultry has never yet risen in price beyond the proportion of other articles of food, and since the demand of the markets has been supplied in as full a measure as formerly. Suppose a heath or common, on which poultry has been customarily bred, is enclosed and improved into farms, is it not probable that, generally at least, as large a quantity of poultry is reared as upon the land in its former state of waste, and producing no corn, a food so absolutely necessary for that kind of stock? In fact, it is open to the observation of every one, that poultry has never been in this country a favourite or prevailing article of diet with the lower or middling orders of the people; thence our farmers, whether little or great, could never be more profitably employed, whether for themselves or the community, than in the production of the more substantial articles of food: in the mean time, the demand for the luxury of poultry never fails to be satisfied to the utmost extent.

GALLINACEOUS FOWLS. In the opinion of physicians, both ancient and modern, the flesh of the chicken at three months old is the most delicate of all other animal food; thence best adapted to the stomachs of invalids, being free from irritation, and affording mild nourishment. Age makes a striking difference in the flesh of fowls, since, after the age of twelve months, it becomes tougher and more insoluble. The cock, indeed, at that age, is only used for making soup, whilst the pullet is still excellent, although a more substantial viand than the chicken. Whilst young, the cock and hen are equally delicate. The capon, or castrated cock, has ever been esteemed one of the greatest delicacies, preserving the flavour and tenderness of the chicken, with the juicy maturity of age, the flesh yielding a rich and good chyle, and without any tendency to inflammation. Capons, however, are usually *crammed*, and made excessively fat, perhaps to the verge of disease, in which state their flesh is neither so delicately flavoured, nor probably so wholesome, as when more naturally fed. Indeed, the flesh of the " barn-door " fowl, or that fed at the corn-store in the farm-yard, is probably the most delicate, as it is universally acknowledged to excel in genuine richness of flavour, which may be attributed to the full allowance of the finest grain they enjoy, as well as the constant state of health in which they are kept by living in accordance with the laws and requirements of nature, and in full liberty and exercise. There is probably greater variety of size, figure, and appearance in the chicken, than in any other species of fowl, and also considerable variety of quality, which will be pointed out under their different heads. The flesh of poultry can be dished up in a greater diversity of modes than that of almost any other animal, and is esteemed as a delicacy by every class of the community. That poultry should not hitherto have formed a more important article of consumption amongst us, may perhaps be attributed to our beef-eating predilections. Some statistics furnished some few years since to the French Chamber, exhibited the singular fact, that whilst in England one-half of the land was devoted to pasture, or the producing of food for cattle intended for consumption, in France only one-fifth was allocated to that purpose: in the former country, the average annual consumption of butcher's meat by the population was 134 lbs. per head, whilst in the latter the average only reached 48 lbs. It is to be regretted that, to render these interesting statistics complete, the relative consumption of poultry and eggs was not also ascertained; but there can be no doubt that the French consume an immensely greater quantity of that article of food than ourselves. EGGS have always been highly esteemed by man as a delicate, agreeable, and nutritious diet. They are dressed in various ways, and enter into the composition of numerous dishes, and, either alone or in combination with other articles, form at once a palateable and substantial aliment. They are eaten with little regard to any thing besides freshness; and

few persons (unless invalids) perhaps have a sufficiently delicate palate to observe the difference of
flavour which exists—eggs varying in size and
quality almost as much as poultry: the eggs of
pullets or young fowls are never so heavy or rich
as when the fowls are mature; those laid by the
larger breeds are considered less delicately flavoured; large eggs have usually pale yolks, and
are of inferior flavour, whilst small ones have
bright deep-coloured yolks, and are more rich,
and possess a good flavour; dark-coloured shells are
generally supposed to have the richest yolks, and
having a stronger flavour are more suitable for cu-
linary purposes; eggs with pure white shells contain more albumen, and are much more generally
preferred for boiling. It is needless to state the
various uses to which eggs are applied, not only
fresh, but also preserved; in the latter state they
are employed on ship-board in long sea voyages,
either as an article of food, or as a substitute for
milk. Every one, however, relishes and enjoys a
newly-laid egg: and it is perhaps for this very
reason that eggs are now seldom or never hawked
about the metropolis, as the Londoners regard the
" warranted fresh " with some suspicion. It is
also much to be regretted, that in London (and
indeed most large towns) the prices of eggs and
poultry preclude their appearance upon the tables
of any but the more affluent classes: if those articles were more readily and generally obtainable

by the residents of cities and town districts, (which they might very easily be rendered, by increased attention to poultry-keeping as a part of the rural and cottage economy of the country,) not only would our home produce take the place of foreign importations in our daily consumptions, but the quantity eaten would be largely increased. By our forefathers, eggs were held in a sort of half religious veneration, chiefly as symbolic of the world and the four elements: the shell represented the earth, beneath it was the air, whilst the white constituted water, and the yolk fire. Various superstitious observances also attached to the egg: the *first* or "virgin egg" of a pullet was the shepherd's gift of good luck to his mistress; the *last*, and that laid on Good Friday, were charms for the well-doing of the whole poultry yard; or if given to a baby taken out for the first time, it brought unfailing "good luck" through life: we are also assured in the *Discorerie of Witchcrafte*, 1665, that to "hang an egg laid on Ascension Day in the roof of a house, preserveth the same from hurts." The agricultural value of the DUNG of fowls has been too long overlooked: "the most valuable fertilizer that we have (says an American writer) is poultry manure, and it is really lamentable to see what a waste is going on around us in this country of the richest and most valuable manure ever discovered: we are importing ship-load after ship-load of Guano, (nothing but sea-bird manure,) whilst hundreds and thousands of tons of our home-manufactured poultry guano, which is said to be equal in value to the foreign, is suffered to go to waste,—each farmer's poultry yard producing so little that it is considered to be not worth bestowing a single thought upon it, and by that means an absolutely large money value is annually lost to the country." It is mentioned that a ten-acre piece of land manured with the dung of fowls, produced no less than an average crop of sixty bushels of corn per acre. Our farmers and small agriculturists will do well to ponder these very sensibly written remarks, as they are certainly worthy of their consideration and attention.

Cochin China.—In judging of the relative qualities of fowls, it must be evident that their chief value must be measured by their powers of producing eggs; and certainly in this the Cochin China far excels all other sorts, in regard to the greater number they lay in the season, and also (which is more important and useful) during the winter, when such articles are always more required and valuable, and generally exceedingly scarce: indeed they seem to possess the invaluable quality of producing an abundant supply of eggs when they arrive at a certain age, (varying from five to seven months accordingly as they have been reared under favourable circumstances or not,) quite irrespective of the season or wea-

ther—contingencies by which the laying properties of most other fowls seem to be very materially
affected; and they furthermore would appear to
retain this good qualification to the very last.   It
is also well established, that they lay much earlier
after hatching a brood, or after undergoing the
process of moulting, than any other kind of poultry: under the former circumstances, hens have
been known to commence depositing eggs in the
nest' within the space of 14 days after a young
brood has been hatched; and during the winter
of 1852, a friend having seven Cochin China hens,
obtained from them, without any extraordinarily
good keep, forty eggs during the Christmas week.
Although undeniably good "brooders," they usually
lay from 25 to 40 eggs before they exhibit any desire to incubate their produce,—but some have laid
upwards of 100 eggs before showing such signs.
Although the term " every-day-layers " is already
monopolized by one or two of our old-established
kinds, the Cochin variety seem even better entitled
to it, as the majority of these fowls will, with ordinary feeding and attention, generally produce
five or six eggs during the week: it is quite true
that in a morbid state of the egg-organs they occasionally deposit two and even three eggs during
the twenty-four hours; but the absurd attempt
made by some writers, on their first appearance,
to induce a belief that such productive powers are
the characteristic of the variety, has long since

been exploded, and scarcely deserves a parting
notice. The egg is not large, 'seldom exceeding
2¼ ounces in weight, and even ranging under;
the shape is usually very rounded, and the shell,
which is of considerable thickness and much granulated, is easily distinguished by the buff colour,
more or less deep, covered with very minute white
specks, generally resembling the egg of the silver
pheasant: though well flavoured, it is not so rich
as the popular belief in that quality being coinci-
dent with a dark shade of shell, would induce one
to imagine they would. They are remarkably
good "sitters," very steady, and cover even comfortably a much greater number of eggs than the
common fowl. In spite of the encomiums bestowed
on them by some of their patrons, the balance of
evidence elicited by our inquiries establish the
fact, that as *table* birds the Cochin Chinas are ungainly; and the quantity of offal in the neck, legs,
&c., the *comparative* leanness of the breast, and
rather pheasant-like or "gamey" flavour of the flesh,
(sometimes also dark, coarse, and stringy,) render
them undesirable for that purpose; whilst the un-
inviting hue of the skin precludes their appearance
as a boiled dish: on this point much diversity of
opinion exists, as some high authorities assure us,
that a roasted Cochin is most juicy, short, and delicious eating: undoubtedly high feeding and superior breeding does produce fowls with flesh both
white, tender, and delicate, and it is therefore dif-

ficult to say what care and attention in those particulars may be able to accomplish. One advantage in this breed, at least to those amateurs who have only a very limited space to allot to poultry conveniences, may be stated, they will bear confinement much better than other kinds—not that we mean to imply that they (any more than other fowls) *thrive* in confinement within such circumscribed limits, but it may be more properly said, that they are injured less, or suffer less from it.

Brahma Pootra.—We have not sufficient experience of this rare and beautiful breed, to be enabled to describe, so fully as it deserves, its characteristic qualities. In the United States of America they have the credit of surpassing (if possible) even the Cochin China themselves as egg-producers; whilst considered as dead poultry they are without question better covered with meat, have comparatively less offal, and are in no way less delicate eating. Dr. Bennett says " as layers they are unsurpassed by any; . . . . they lay a larger egg than any other Asiatic breed with which I am acquainted, and I have tried them side by side with the Cochins, and find them equally prolific." Mr. Miner, the editor of the *Northern Farmer* (U. S.), says, " we presume there are no other domestic fowls that lay a larger egg than do the Brahma Pootras: and they certainly equal the very best sorts as regard laying. According to Mr. Nolan, the egg was as large as the Turkey's, but we have never seen any of such proportions : they are, however, larger than those of the Cochin China, have a lighter coloured shell, and are equally well flavoured."

Malay.—Fowls of this variety, although not superior, are at least plentiful layers : if well fed, they produce eggs of a very fair size, of superior richness and delicacy to the Cochin, and affording also more substantial nutriment, and they have further the character for being superior, as regards vital productiveness in hatching, to all other fowls' eggs : the shell of the genuine Malay will, we think, be found invariably of a buff shade, and not, as it has been asserted, " perfectly white." During the laying season, (which, however, does not last so long with them,) they almost rival the Cochin China in the number of eggs they produce : some Malay hens, belonging to Mr. I. C. Dowsing, Great Yarmouth, during the whole of January, 1853, did not miss laying more than four days, and in the three succeeding months the relative number of eggs produced preserved the same relative proportion. Some have thought, from their long legs, that they cannot generally be good or steady sitters, but a good deal of prejudice against them exists, as those with whom they are favourites maintain that they are fair sitters and tolerably steady. The length of legs and body, and want of compactness, render them undesirable as stock for the table, and their flesh is usually dark and coarse, more adapted for making soup than eating, although if well fed the flesh is good flavoured, notwithstanding the yellow tinge which usually pervades the skin. Like the Cochin China, they possess the recommendation of submitting quietly to a confinement in a very limited space—often rendered a very necessary infliction, (even when not compulsory from want of convenience,) owing to their rather quarrelsome and pugnacious conduct towards other fowls.

Game.—Fowls of this variety are no longer valued for their fighting qualities, and they must even condescend to be tested by the same rigid standard of practical merit as other fowls,—unless indeed we regard them as ornamental birds only, when to our mind this would be the variety which of all others the amateur ought to select. Tested however for economic qualities, they cannot be said to be prolific layers of eggs, on the contrary, they seldom present their owner with an egg oftener than every alternate day,—and that too of a size even somewhat smaller than the Cochins, though finely shaped and extremely delicate: but as steady sitters and good nurses they are never beaten and but seldom equalled. For the table the Game fowls are neither large nor fat, but their flesh is of the most beautiful whiteness, and superior to that of all other breeds of domestic fowls for the fineness of its fibre, and the richness and delicacy of its flavour, for which it is highly prized by epicures of such meat. They require a wide and extensive range, and this coupled with a naturally cruel and pugnacious disposition, has entailed their banishment from almost every yard, except such as are exclusively devoted to them, by amateurs or breeders who still admire them for their genuine and characteristic English " pluck," (to use a vulgar though expressive word,) which induces them to give battle at all times and to " all comers."—The *Shake-bag*, which, though a cross-bred fowl, ought not perhaps to be quite passed over, and may therefore have a place conveniently assigned it here,—was formerly celebrated for the fineness and whiteness of its flesh ; and from its size it formed an excellent substitute for a turkey—to the frequent great convenience and accommodation of the poulterers and innkeepers of the town of Wokingham and other places.

Dorking.—These far-famed fowls cannot certainly be said to be either extraordinary or surpassing layers, but they undoubtedly furnish a supply of eggs tolerably plentifully, and of good size, though not so large as the superior proportions of the breed would induce one to believe would be the case. It is quite true that occasionally some specimens almost rival the Cochin China in productiveness, but generally this quality has been much over-stated as regards them, and as a consequence many amateurs have been greatly disappointed in their expectations upon this head :

they will usually, according to the season, each lay from 140 to 175 eggs, in addition to rearing two broods of young; the eggs have a pure white shell, and are peculiarly rich and well flavoured: it is said that the *white*, or original Dorking, is the most prolific layer in summer, whilst on the contrary the dark-coloured, or so-called " Improved Sussex," breed excels for winter laying. As steady sitters and good nurses they are of surpassing excellence—equalling the Game variety for the former quality, and outstripping it in the latter; nor is there any thing injurious or inconvenient in the fifth claw or toe which distinguishes them,—that supernumerary member being seldom of a sufficient size to encumber the foot, or to cause it to scratch out its eggs from the nest, as some have pretended is often the case. An opinion prevails that when *young* the Dorkings are much less hardy than the Cochins and Spanish fowls; but this may be a popular prejudice, as many breeders have assured us, they have not experienced any extraordinary trouble or difficulty in rearing them. For the table uses the Dorking is unquestionably superior to all other breeds; for although it may not exceed the Game fowl in delicacy, it yet surpasses it in the quantity of meat it affords, and that too of beautiful whiteness and richness of flavour: the pure *white* Dorking is said to possess a yellow or ivory hue in the skin, most undesirable in any fowl,—at least this fact is noticed by Mr. Baily, an authority we cannot dispute, although our own observation does not confirm or establish it. From its peculiar adaptation to the uses of the table, this is the kind most generally made capons of—indeed the qualities of the Dorking may be emphatically stated to constitute it essentially a *flesh-producing* rather than an *egg-manufacturing* fowl.

Spanish.—The superior excellence of this variety is so generally known and appreciated, that it almost seems superfluous to speak of their merits as layers; in which capacity they have long been highly valued as very abundant producers of fine, large, well-flavoured eggs of most unusual weight, though some may think they are not quite so delicate eating as those of the smaller varieties; the eggs will average from $3\frac{1}{4}$ to $3\frac{1}{2}$ ounces each; and good hens will generally lay four or five days in every week from about the end of January to September or the moulting season, which however is a long one with them, and during nearly the whole of winter, one or two eggs weekly will exceed what can fairly be expected from them: they commence laying at about six months old— very nearly as early as the Cochin China. During the month of June, 1853, one Spanish pullet and three hens laid 92 eggs, being an average of 23 eggs each in the thirty days, and that too without any extraordinary high feeding. What adds to their value as layers is, that they seldom or never exhibit the slightest disposition to incubate,—although this may be regarded by some as a disadvantage in the variety, at least when stock is required, for they must necessarily have the character (as may be readily inferred) of being bad sitters and nurses. The flesh is both white and delicate, and of good quality, but generally speaking they have too much offal, and are not adapted for fattening well, except as capons,—they cannot, therefore, be considered as ranking at all high as a table fowl.

Poland.—These are not alone kept as ornamental fowls, but they are certainly one of the most useful varieties we have, more particularly on account of the abundance of eggs they lay, indeed it is generally conceded that they are remarkably good layers. It is thought that the White-crested Black Poland is the best egg-producer, as far as regards the actual *number*, whilst the spangled varieties lay *heavier* eggs: the produce is about equal to the Spanish numerically, but not in weight. The Poland begins to lay early in March tolerably abundantly, and continues to the moulting season with great regularity. An average produce during the year will probably be between 130 and 200, but we have heard of instances in which as many as 233 eggs have been obtained from one Poland fowl: the egg weighs about $1\frac{3}{4}$ ounce, and has a fine white shell. From October 25th in one year, to the 25th September in the following, (eleven months,) five Poland hens produced 503 eggs, one only sitting during that time, and the whole being kept on ordinarily good feed merely. Another instance occurred in which a Poland hen was mated with a cock on the 7th December; she commenced laying on the 28th of the same month, and by the 1st of March following had produced 56 eggs. The bearded spangled Poland will, we think, be found very little inferior to the black as layers, although popular opinion is opposed to this view: Dr. Horner informs us that one hen in his yard commenced laying in March, 1852, and never ceased until the end of September, depositing an egg for two successive days and then missing, giving an average of five eggs per week during the season; that gentleman adds, "none but the true Sonnerat's fowl are so admired or are so strong, and they are unequalled layers; they never sit, at least the true-bred bearded Polands do not." The black occasionally evince some desire to incubate towards the end of summer; they are not, however, to be depended on, as they are very bad, unsteady sitters, even the best of them being less inclined than almost any other fowl. As articles for table use they are excellent, for, although not large fowls, they are plump and fat, the meat being delicately flavoured and of good colour, equal in that respect to the Dorking, and perhaps more

z

rich and juicy: they fatten as quickly as any breed.

Hamburgh.—The old name by which these were popularly known, " Dutch *every-day layers*," sufficiently expresses the superiority of their producing capabilities—which really seem " everlasting." Mr. Bailey justly describes them as " prodigious layers naturally, and without over-feeding." They commence their task very early, sometimes at four (seldom delaying it beyond five) months, and if kept up in good condition, they will maintain a regular supply during the whole year, except perhaps a brief time when on the moult: the egg is small, from $1\frac{1}{2}$ to $1\frac{3}{4}$ ounce in weight. A careful trial of three months with four pencilled Hamburghs, gave an average (collectively) of 26 eggs weekly: and instances have been communicated to us where the annual supply from the same varieties has averaged from 230 to 256 eggs. Their laying is never interrupted by sitting, such an occurrence being exceedingly rare—indeed almost unknown. We should place the *pencilled* fowls in the first rank as layers; whilst for the table, though small, they are well covered with meat at once white and tender. The *spangled* yields a greater quantity of flesh, and eggs also of a larger size though fewer in number.

Bakies.—The comparatively recent introduction of this valuable variety has prevented any

extended observations upon their merits or qualities: and we are indebted to John Fairlie, Esq.,
(the principal possessor in this country,) for the
following information as to their merits. "They
are most excellent layers, and certainly surpass
the Dorkings in the number of eggs they yield—
they often lay a nest full, (if the eggs are regularly
taken out,) and then, taking to another, will lay
enough to fill that also, before they attempt to sit;
which when they do, they fully justify the oft-re-
peated remark made upon them, ' *What excellent
sitters they must make!* ' They cover many
more eggs comfortably than one would suppose
from their size: on the nest they appear as if they
were squeezed quite flat down: they are very
gentle when hatching, and are good and assiduous
mothers, though their legs are so short that as
they stand up the young can nestle under them;
the hens are also the best I ever knew to take to
the chickens of other fowls. The egg has a much
whiter shell, and considerably larger than those
which generally find their way to our markets.
As dead poultry their bodies are plump, and the
flesh very similar to that of the Dorking,—in fine,
altogether it is a most valuable and useful
variety."

Bantams.—Although almost exclusively regarded as fancy stock, they yet perhaps possess
sufficient good economic qualities to deserve to be
rescued from the silence to which it is our pur-

pose to assign the Barn-door and other nondescript varieties. They are tolerably good layers,
but of such miserably small size, as very little to
exceed a large pigeon's egg: during the season
from the beginning of March to the beginning of
October, a supply of from 130 to 150 eggs may
be reckoned on from any good Bantam hen: if
high-bred, a winter supply may sometimes be obtained,—in one case upwards of 185 eggs were
laid by six little Bantams from January 1st to
March 31st, in a season by no means mild. Bantams are capital sitters, and for hatching purposes
are particularly useful for covering the eggs of
Pheasants, Partridges, &c., being also good nurses.
For the table, from their size and delicacy they are
very convenient, as they may always stand in the
place of chickens, when small and young ones are
not otherwise to be obtained; and as a substitute
for partridges they are not to be despised. They
are great devourers of wood-lice and other insects
so destructive to plants, and therefore, but for
their inveterate habit of scratching every inch of
ground over, the Bantams would be invaluable—
as it is, the good would perhaps be more than
counterbalanced by new and more vexatious floral
spoliations.

Nondescript, &c.—Including as we do the Silky
Fowl, and other anomalous birds, under this head,
they can only be regarded as fancy poultry, and
no one would dream of stocking his yards with

such for their economic value, either as the means of obtaining a daily supply of eggs, or of afterwards placing upon the table.

Pheasants.—This fowl is not to be tamed by domestication like others; nor is the flesh of those which are brought up in the house in any degree comparable with that of the wild Pheasant, or that living in a state of nature: and for this reason, when bred at home, it is either merely as objects of show, or for the purpose of replenishing the proprietor's grounds. In a wild state the hen lays from 18 to 20 eggs in the season, but when subjected to confinement very seldom more than ten. The wholesomeness of the flesh of the pheasant was proverbial among the old physicians; it is of high flavour and tolerably digestible, being in greatest perfection during the autumn: a young bird very fat is reckoned an exquisite dainty. Although very great eaters, they are not altogether profitless, as it is well known they greedily devour minute insects and vermin: some years since, a pheasant was shot by T. Day, Esq., of Herts., in the crop of which was found contained more than half a pint of that insect so destructive to farm produce—the wireworm.

Peacock.—This bird has long ceased to form a common dish at the table—if indeed it can ever be said to have done so; and probably, from the usual coarseness and ill flavour of the flesh, the motive for introducing them at banquets was always rather for show than use: Pea-hens and Pea-chicks, however still retain (occasionally) their place at fashionable tables—just as they were in the days of old Gervase Markham, "more for rarenesse than nouryshment:" as poults or chicks they are more eatable, and even when older, if well fed or fattened, they are not to be slighted. The Pea-hen lays but very few eggs in a season— five or six, though if removed from her she will sometimes be tempted to repeat the dose: the egg is larger than that laid by the Turkey, and the shell is of a slightly speckled creamy-white. The Pea-fowl is very unprolific, and quite unprofitable as a table fowl, and of course can only be regarded as an ornamental bird for the lawn or plantation, as they are from their selfish, vicious, and destructive habits most undesirable appendages in the poultry yard. Exclusive of the consideration of ornament, the peacock is very useful for the destruction of all kinds of reptiles; but at the same time some peacocks are said to be so vicious as to tear to pieces and devour young chicks and ducklings, suffered to be within their reach. They are also very destructive in a garden, committing wholesale depredations upon both buds and fruits.

Turkey.—Although the Turkey hen lays a considerable number of eggs, varying from 12 to 22 at one laying-time, (16 or 17 being the usual number,) yet they do not produce enough, nor are the eggs sufficiently valuable, to enable them to become an article of consumption: during the season the hen lays sometimes daily and sometimes on alternate days until a "sitting" has been deposited: the eggs are large and rich for household purposes, but are of more value for hatching; the shell is somewhat spotted like that of the Guinea Fowl, but the wild Turkey's is said to be more of a clear bluish white. The hen is a remarkably steady sitter, so much so that it is sometimes necessary, whilst incubating, to force her off the nest to prevent her starving. The Turkey is a fowl which is now so generally reared and consumed, that most persons can appreciate the delicious viand it places on our tables. The genuine "Black Norfolk" breed (so called) is generally said to possess flesh of a superior quality to any other—at least it has that reputation, although there are some cooks who prefer that of the White. The flesh of the Turkey is somewhat more dense of fibre, and more savoury and substantial, than that of the chicken, but it is reckoned nourishing and restorative. Age produces a similar effect as in the chicken, whence the Turkey, after a certain period, is good for little, except stewed. The Turkey has ever been remarked for its fulness and weight of flesh in the breast—no doubt, beside, the prime part. The dead weight of a fat Turkey being 21 lbs., according to the late Mr. Young, renders 14 lbs. when ready for the spit.

Guinea Fowl.—Although very unpopular birds among poultry keepers, Guinea Fowls are most prolific layers, yielding an abundance of good and nourishing eggs. Indeed the statement of the Rev. E. S. Dixon, that " the body of the Guinea hen is a most admirable machine for producing eggs out of insects, vegetables, grain, garbage, or whatever an omnivorous can lay hold of," has more of truth in it than the humorous conceit would imply: a Guinea hen has been known to lay as many as 73 eggs in seventy-five days, and without attempting to search out a new nest, although the eggs were removed daily,—but this, it must be admitted, is a remarkable instance; the average produce, however, of the season from March to the moulting period, (usually in September or October,) will be found to vary between 170 to 190 eggs, according to the care, attention, and food bestowed upon them: the eggs are small—less in size even than those of the Cochin China—but they are not the less rich in quality, containing *comparatively* a far greater proportion of yolk, of a deep orange (almost red) colour, generally of a rather high flavour or wild taste, though in this latter point the eggs are more materially affected by the nature of the feed, than those of any other fowl: the shell is of great hardness, and of a deep buff or light chocolate shade and speckled. They are fair sitters when once they undertake the task of incubation, but never exhibit any very marked or strong desire for that occupation. As table delicacies their flesh is highly esteemed by those who have tasted it; the flavour is high, and has much of the wild or "gamey" taste, and in consequence they are reckoned by some an excellent substitute, especially as they are in season about the time when game is going out, namely, from the beginning of January to the end of May. They are not so white of flesh as the common, but more inclined to the pheasant colour; full breasted, and in quality firm, short, and savoury like the flesh of the Pheasant, and easy of digestion. The Guinea Fowl is said to unite the properties and characters of the Turkey and Pheasant,—it is courageous, active, inclined to be pugnacious, and of a restless, roving disposition; however long domesticated, these birds retain some part of their original wild habits, and will stray in search of a place in which to drop their eggs, without any apparent solicitude as to their security, and accordingly require much attention and management: they are also extremely destructive, and commit frightful ravages in the garden, where they gorge themselves on the choicest productions—indiscriminately attacking fruits, vegetables, seeds, roots, &c.

PIGEON.—Although the varieties included in this class more particularly appertain to what has been very justly termed "the Fancy," still the commoner sorts often make a welcome appearance upon our tables in the shape of pies or other equally agreeable forms,—the only economic uses to which they can be applied, as the small size and limited number of the eggs render them valueless except for hatching purposes, which they will perform on six or eight occasions during the year, each time upon a pair of eggs. The flesh of the wild Wood Pigeon is in perfection in the latter summer and the autumnal months, from their ability in those seasons to obtain the best food; during winter, feeding on coleworts or any green food they can find, their flesh is loose and bitter: from their large size, which would be increased by domestication, the experiment might be worth the trial, of bringing them up in the cote. The flesh of the Wild Blue Rock when young is esteemed the most delicate of all. Among our domesticated breeds the common " Dove-house " kind are unequalled as mere breeders or producers of stock; but we think the Runts are preferable, from their superior size, and the ease and rapidity with which they may be fattened; they likewise breed well and tolerably fast, and to our fancy are the only sort that can be sent to the spit without being spoiled: as producers of ingredients for our table dishes, the Nuns, Antwerp Carriers, and Turbits, should perhaps rank next, but their present great value as Fancy varieties would no doubt prevent their sacrifice to the kitchen,—but generally speaking no other fancy sorts can be regarded as profitable stock. A great many absurd conceits are found in old books relative to the quality of the flesh, and to the medicinal and remedial properties of almost every part of the pigeon,—" they are (saith Willoughby) far harder to concoct [digest] than chickens, and withal yield a melancholy juyce:" little of such gossip can be relied on. Their flesh, when young and in good condition, is both a nourishing and stimulative diet; that of the full-grown pigeon more substantial, but harder of digestion, and in a degree of a heating nature. As a general rule, with pigeons at least, the dark feathered are thought to have rather dark flesh, of high flavour, and with the gamey bitter of the wild birds; whilst the lighter feathered have flesh of a proportionately lighter colour and more delicate in flavour. Shooting matches constitute another sport to which these birds are applied, and of which periodical details are to be found in the newspapers. Their dung possesses an extremely heating and drying quality, and hence is highly valued as a manure by such agriculturalists as are conversant with its uses and nature; it was also formerly applied medicinally in various diseases. In the East these birds are extensively kept merely for the sake of the dung: in Ispahan, according to Tavernier, there are no less than 3000 pigeon-houses, nearly all belonging to the king, who derives a very large revenue from the sale of the manure, which is employed in the cultivation of melons: these eastern cotes are described as being " large round towers, broader at the bottom than the top, which is crowned with conical spiracles through which the pigeons descend—the interior being pierced with a thousand holes, bearing a close resemblance to honey-comb."

SWAN.—The Cygnet, or young Swan, only is reckoned eatable, and that after a peculiar preparation, although in old time the Swan formed a dish of embellishment and show at great feasts. The fat possesses probably much the same qualities as that of the goose, but is more mild and emollient. Many curative virtues were attributed by the ancients to the Swan's skin, but modern practice only sanctions its use as a defence against rheumatic affections; in fact, the only worth of the very few wild Swans which reach our market, consists in their skins, which are sold to the furriers. Exclusive of ornament, the chief use of the Swan is to clear pieces of water from weeds, a service effected some years since by those birds over a very considerable breadth of water at Clumber, the residence of the Duke of Newcastle, in the course of a year or two: but they are generally reputed great destroyers of the young fry of fish. There is a sheet of water at Burghley, about a mile in length, which used to be so over-run with weeds, that three men were constantly employed six months in every year to keep them under; two pairs of Swans completely cleared away the weeds the first year, and none have since appeared, as they eat them before they rise to the surface. In the Low Countries, Swans are kept for the same purpose. The Swan will lay generally from 6 to 10 eggs, which are of course used only for hatching: they have a hard, white, and somewhat stuberous shell. A fat Cygnet is highly prized by epicures: they are in season from June to December, but the best time to eat them is in October and November: old birds are usually made into soup, or boiled.

Goose.—It is a curious illustration of the *de gustibus non est disputandum*, that the ancients considered the Swan as a high delicacy, and abstained from the flesh of the Goose as impure and indigestible; whilst the moderns generally now reject the flesh of the Swan, and eat that of the Goose with an universal relish. The national antipathy to the flesh of the Goose, which Cæsar recorded some eighteen centuries ago to exist among the Britons, had not altogether disappeared from some parts of this kingdom even so late as the last century: thus in Cornwall, the peasants, like our old Druidical ancestors, had the greatest prejudice against the flesh of what they termed " hollow fowl," under which they included hares, chickens, and geese—the " *leporem, gallinam, anserem,*" of Cæsar. The whole *anserine* tribe, however, are now held in high repute by lovers of good eating, and are allowed to afford a highly stimulant food, of a strong flavour, and viscous quality, and of a putrescent tendency. The flesh of the domesticated Goose is more tender than that of the wild; but it is a diet only adapted for good stomachs and powerful digestions, owing to the flesh being permeated with fatty substance, rendering it more difficult of digestion than that of other kinds of fowl; indeed an over-fattened Goose is too much in the oil-cake and grease-tub style to admit even the idea of delicacy, tenderness, firmness, or true flavour; the fat is more subtle and resolvent than pork-lard, and is an excellent article to be reserved for various domestic uses. Every body knows that the only orthodox way of eating Goose is roasted, but it is not generally known that the fruit which forms the sauce so inseparably connected with roast-goose, received the name of *Gooseberry* from the old practice of eating it with that fowl. Most of the foreign varieties of Geese are kept simply as ornamental poultry for the lawn, or as objects of curiosity. The common domestic Geese are proverbially very early layers,—every one remembers the old rhyme that tells us " on Candlemas Day," or the 2nd February,

"Good Housewifes' Geese lay;"

and by the festival of St. Valentine, laying is the order of the day with these fowls generally : they usually lay an egg every other day until from 9 to 15 are deposited, when, if allowed, they will incubate; but if the eggs are removed from the nest, and high feeding resorted to, they will lay from 40 to 50 during the season. Some years since, Mr. W. Holmes, of Spaldington Lanes, near Howden, Yorkshire, had a Goose in his possession which, within twelve months, laid 70 eggs,— namely, 26 at the usual time for incubating, (from which she hatched and brought up seventeen fine goslings,) and then she began laying again at the end of harvest, and continued to the close of the year, yielding an egg every other day. Instances are recorded of Geese laying upwards of 100 eggs during the year, as their individual produce. The egg is very large, weighing from five to six ounces, and though rich enough for pastry uses, (if like the Turkey they were not more valuable. for hatching,) are too strong for using *au naturel*. Geese are very steady sitters, commonly rearing two broods in the year, and not unfrequently three: in November, 1853, a Goose belonging to Mrs. G. Fawcett, of Appleton-le-moors Mill, hatched a *third* brood of eight, having previously in that year raised nine out of her first lay of 10 eggs, and nine birds out of the same number of eggs of her second lay. The dung of Geese is very valuable as a powerful manure, and a large flock would have a considerable effect in fining and improving the grass of coarse meadow land;

but in large quantities it is said to be too heating and corrosive for farm land generally, but that is no argument against its use—only against its abuse. Geese attain almost an incredible age, if allowed to live. Many years ago the author was cognizant of one on a farm in Scotland, of the clearly ascertained age of 81 years, which then died a violent death,—her fecundity seemed permanent: other cases have been known of Geese living to 70 years. The following varieties are all requiring separate notices.

Embden Goose.—This has an advantage over most other kinds, in regard to its value for the table, from its larger size, as well as the clean white appearance of the flesh when dead: its flavour is equal to that of any other; but to countervail the increased size, the extra quantity of offal weight, owing to the long neck and legs, must not be lost sight of. They are fair layers of eggs, good incubators, very quiet, never stray, and above all fatten easily.

Toulouse.—Highly esteemed for their large compact forms; and for the table they may be fattened even to repletion without becoming, we are assured, " so disgustingly fat as the other sorts." They are very superior egg-producers, laying unusually early, and most abundantly: their only drawback is, that they are very indifferent incubators and mothers.

China Goose.—Considered but little inferior to

the preceding in its laying qualities; they lay both remarkably early, and also late in the year, and are steady sitters, although apparently not very fond of the task: they will hatch a brood often in January, which may be readily fattened by the end of May fit for the table. Of their qualities for the table we cannot speak. The eggs are smaller than those of the common domestic Goose.

Duck.—Although almost all kinds of wild aquatic fowl supply a most dainty and scarce delicacy for the table, yet they principally serve either to fill the sportsman's bag with trophies, or to store the ornithologist's museum with curiosities: and it is therefore with the Duck in a domestic state that we have to do. The Duck, equally with the Goose, was a great favourite with some nations of antiquity, and deservedly so, for the mildness and simplicity of their character, from their great fecundity, and from the cheapness and ease with which they were provided. The inoffensive and harmless character is common to both species, rendering them most pleasant as well as profitable animals to keep, and comparison between them and the chicken, in this latter particular, is certainly in favour of them. Upon the excellence of the Duck all parties seem to have agreed. The ancients went even beyond our greatest modern epicures, in their high esteem for the flesh of the Duck, not only assigning thereto

2 A

the most exquisite flavour and delicacy, but also attributing to it important medicinal properties; for Plutarch asserts that Cato preserved his whole household in health, by dieting them with ducks' flesh as a prophylactic; surely a most pleasant mode of taking physic! The flesh of the Duck is mostly admired for the high savoury flavour of the brown meat, although it must be confessed that the duckling usually provided for our Christmas markets is a sufficiently insipid viand: that of the fatted fowl is said to afford a preferable and more nutritious diet than that of the Goose, being less gross and more digestible. As we have said, the Duck possesses great fecundity, producing a very large number of full-sized eggs, rich, and capital for use in pastry: the common variety lay about the beginning of December, and will continue with few interruptions till moulting-time in May, resuming that occupation in the latter summer; the number of eggs laid varies very much, from about 60 to 90 in the twelve months, but superior specimens occasionally yield 100 and upwards: the eggs are generally produced every other day, though they will sometimes be laid with the greatest regularity for 40 or 50 successive days; a duck of the common kind, belonging to Mr. John Morrel, of Belper Dally, performed this feat for 85 days. The colour of the shell varies from a dull white to a greenish tinge, eggs with white shells being less strong in flavour: Ducks with white or light-coloured plumage produce the light-shelled eggs, whilst the brown and dark feathered usually give dark shells—a fact worthy of being tested by more extended observation than we have had at our command.

Rouen.—This Duck has rather a darker flesh and more savoury than the common English kinds, and there is much contradictory opinion as to the quality of the flesh, some contending that they are both strong and coarse, whilst others say they are surpassing fowls for the table; we are inclined to think they possess moderate excellence in that respect. As layers they are prolific, and tolerably early, beginning often in January, and producing regularly 3 or 4 eggs a week till they moult; and about September they renew their supply, and do not cease until the intense cold of winter sets in: three of this variety, belonging to Mr. Punchard, laid upwards of 340 eggs during the season from February to July, and of these one bird laid the extraordinary number of 92 in as many consecutive days. The Rouen Ducks are good sitters, but from the close proximity of the abdominal pouch, some breeders assert they are apt to squat too heavily upon their young brood, and therefore do not make good nurses.

Aylesbury.—These are acknowledged to be among the earliest and most abundant of layers, giving from 80 to 130 eggs in the course of the year; the eggs are of a good size, well flavoured, and neat white shell: they are also good incubators, generally rearing two broods in the year, if allowed to sit. The flesh of the Aylesbury is excellent, it is of a lighter colour, more delicate, and of a less high savour, than that of any other breed. They are profitable as poultry in every way; but it is not alone as useful fowls that they are prized, for as ornamental on a piece of water they match prettily with Swans or Embden Geese.

Muscovite.—These and other foreign importations of the Duck family, are kept more out of curiosity than for their economic qualities. Although some writers assert that they are " profitable for their fecundity, size, and aptitude for fattening easily," our own conviction is, and general experience of them will bear us out in saying, —that they are in every way inferior to the Rouen or the Aylesbury variety, and upon the whole cannot but be looked upon as unprofitable stock: they will seldom yield more than from 12 to 20 eggs at one time, and never yield this limited supply oftener than three or at most four times in the twelve months; they are very dirty in their habits, as, owing to their feathers being pervious to wet, they seem as cordially to detest water as their tribe is proverbially fond of it; they scarcely deserve a better place than a corner in a stable yard or piggery, as from their not straying, or having any desire to, they are content within almost any limits,—but they are unmerciful tyrants to all other fowls, and often even to their own young broods: the Rev. E. S. Dixon's summary of their qualities is still less flattering than our own,— " they are gross, obscene, filthy, quarrelsome, and tyrannical ! " With all this, the flesh is also very inferior, coarse, and tough ; but nothing is more absurd than the threadbare statement of most writers, that the meat has a " *strong musk flavour*," arising from an oily rump-gland which must be cut off as soon as the fowl is killed, to lessen as much as possible the disagreeable taste,—for although no doubt the old birds are tough and rank-flavoured enough, yet if eaten when *young*, and scarcely completely feathered, they will be found a passable dish, unless indeed the feed has been bad, when it is difficult to say what ill-savour they will not imbibe.

East Indian.—These possess every quality of a semi-domesticated variety, they are wild in habits, the eggs are wild in flavour, and so is the flesh. They lay tolerably abundantly, the eggs having not only the peculiarity of colour in the shell we have already noticed, but even the yolks seem somewhat different to those of other Ducks, being of a remarkable greenish bilious hue, but otherwise delicate ; and the wild " gamey " flavour of the flesh is esteemed by some epicures, as it is said to provide a fair substitute for wild fowl, when such dainty dishes cannot be obtained or are out of season. They are much more aquatic in their taste and habits than other Ducks.

Rotterdam or *Dutch.*—This variety, if such it may be regarded, is but little known, but our friend, Mr. James Kemp, of Great Yarmouth, who kept them for many years, assures us they are invaluable, whether as egg-producers, incubators, mothers, or as table fowls, far surpassing any other variety he ever had: we are sorry we have no means of recording the number of eggs laid by them in the year, but, on the same authority, we can give some evidence of their extraordinary powers and excellence as sitters; upon two different occasions a duck (not by any means a large one) of this variety laid 23 eggs during the month of October, and *each time* she sat so steadily and closely over them, that she succeeded in hatching and rearing twenty-two young ducklings—such success we believe being rarely attained by any other kind.

SELECTION OF STOCK.—On commencing the pursuit of poultry keeping, the point first to be determined is the object which is sought to be attained,—whether it is proposed to keep and rear fowls simply as ornamental birds, for gratification and amusement, or for the economic value of their produce and their uses for the table—in other words, whether for fancy or profit; for under these two divisions may be included all who can be legitimately regarded really as poultry keepers, although, doubtless, there are many persons who take up the occupation, not with any defined purpose or object, but merely because it happens to be just now the prevailing popular mania or fashionable whim. If the former be the object, in the choice of the *variety of fowl*, we can only offer the amateur one piece of advice, namely, *to exercise his own taste and fancy alone*, subject of course to the restriction which a consideration of the conveniences and accommodation available and at his command must necessarily impose: to give any other advice would be useless,—the Game, Poland, Spanish, Hamburgh, and Bantam classes, afford a sufficient diversity of appearance and plumage, and have each their respective admirers; and after all, it is vain to stock a yard with varieties in which the proprietor sees no beauty, and takes no pride or pleasure in tending, merely to be recompensed now and then by hearing some commendatory remarks bestowed upon them by a friend or neighbour, who really, perhaps, knows little and cares less about fowls. But having once decided on the variety he intends to essay to keep, we may, it is hoped, be able to offer some few hints which will prove useful to the amateur in selecting the individual fowls: if he unfortunately cannot decide the points of excellence for himself, even with the aid of our full description of the different varieties, let him by all means obtain the advice and judgment of a friend who does know the difference between good and bad birds; for let him rest assured, to stock his yard with inferior fowls will to a certainty result only in great annoyance, useless expense, and unfailing disappointment and vexation,—an opinion in which we are strengthened by no less an authority than Dr. Horner, who observes to us, " I am convinced by practical experience that the only proper way is to get the *very best of everything at first*, and to reject at once all others that are not *the best:*" and that this is sound advice must be apparent to the beginner, when he reflects that it takes as much time, anxiety, attention, and expense, to rear inferior as it does superior fowls, with the disadvantageous addition that no single good end of any kind is in the former case attainable: if time and money therefore be not a consideration, then let the stock of fancy fowls be selected, in the first instance, from some of the *prize pens* of birds exhibited at the Poultry Shows, now so frequently occurring in almost every district of the kingdom; it is very true that such birds have often affixed to them most absurd sums, far above their value, and which must (as indeed they are intended to) act as altogether prohibitory of any sale; in which case, however, the eggs or young progeny of such may sometimes be procured at reasonable prices, or, which is even better, some " commended " (very slightly, if at all, inferior to prize birds) may be obtained at such exhibitions for sums much less exorbitant or extravagant. If however the amateur resorts to poultry keeping as a source of profit, he will have still another question to decide, as it will necessarily very materially guide him in selecting his first stock,—it remains for him to consider whether eggs or dead poultry are the most valuable and saleable in his immediate district: if eggs are much sought after and fetch fair and good prices, he must select those varieties which experience has proved to be the best producers of those articles at the smallest cost; whilst in the case of dead poultry being the most marketable commodity, clearly his choice should fall upon such as are the most generally esteemed for their table qualities, and which evince a greater readiness to fatten quickly, and at a less expense, than others. In any case, we trust that a reference to the pages preceding,—which have been devoted to a description of the edible qualities and powers of production in the different varieties of our domestic fowls, in connexion with succeeding hints and suggestions, will enable the amateur to make a judicious selection of stock, such as will prove the best calculated to fulfil the requirements for the table and market, or as may be the most suitable to his fancy and situation.

A breeding stock, of the common kind, is easily procurable, either in town or country, from the markets or individuals: particular and fancy breeds can either be sought in those parts where they are customarily bred, at the shops of the London dealers, or from those Fanciers in whose hands they may be.

Eggs, as a *general rule*, are more profitable produce than fowls, for the simple reason that as soon as they are produced (in fact the nearer that time the more valuable are they) they find a ready sale, new-laid eggs being almost invariably sought after in greater quantities than they can be supplied; whereas if chickens are hatched, they require some months of care, trouble, and expense, (if reared at all,) before they can be brought to market and their value realized,—but of course the relative value is often so materially altered by local circumstances, as to render any arbitrary or fixed opinion of no value. In localities where (as in the case of most towns) eggs fetch good prices from being much more than ordinarily in demand, it will manifestly be wise to make choice of fowls which are abundant layers, as the year's proceeds from such produce cannot fail to exceed those derived from sales of chickens for table purposes; but on the other hand, where chickens meet with a ready market in the *immediate neighbourhood* of the proprietor, and at anything like prices which are remunerating for the trouble and cost of rearing, it will be far preferable in every point of view, to fatten the young birds as speedily as possible, in order to dispose of them as dead poultry, for even if the actual sum realized, at the expiration of the time required for fattening, be not more than half what the year's produce of eggs would have brought, still we must then consider, as a set-off, that the cost for food for the young fowls would be little more than one-third of what a twelvemonth's keep of laying fowls will come to, and in addition the risk of loss by disease is to a great extent avoided.

Perhaps the most profitable way of disposing of produce and fowls is through the " higgler " or middle-man; for although it may be said that this plan runs away with a large portion of the profit, still if it is remembered that there is no expense for carriage, no risk or losses, a certain and ready sale, and withal comparatively no trouble,—it is, to say the least, doubtful whether the advantages, attending this mode of getting rid of surplus stock, do not more than countervail the disadvantages of diminished returns.

For breeding-stock it should be almost an undeviating rule to take *young* fowls; a two-year old cock and a hen in her second year, are perhaps the best ages for stock-birds: and we may be excused for giving an explanation of some very simple terms applied frequently to poultry, (and as often incorrectly,) as denoting the various stages of their existence; under the term *chicken* are rightly included young birds of both sexes, but the *age* at which this phrase becomes inapplicable to such, appears not yet determined; popularly, however, the young fowls retain it until about four months old, when the male is called a " cockerel," or " cockling," and the female a " pullet," until they arrive at the end of the first year of their existence, when the former is advanced to the title and dignity of a " cock " or " stag," and the matron to that of " hen."

For stock the most useful cock is generally a bold, active, and savage bird, cruel, arbitrary, and even destructive to his hens in his fits of passion, if not well watched, and sometimes to his own young ones.

Hens are in their prime at three years of age, and decline after five, whence, generally, it is not advantageous to keep them beyond that period, with the exception of those of capital qualifications. Pullets in their first year, if early birds, will probably lay as many eggs as ever after, but the eggs are small, and such young hens are unsteady sitters, and cannot judiciously be bred from for stock. Hens with a large comb, or which crow like the cock, are generally deemed inferior; but we have heard of hens with large rose combs, and crowers, which were upon an equality with the rest of the stock. Old hens are seldom to be depended upon for eggs in the winter, such being scarcely full of feathers until Christmas; and then often will not begin to lay till March or April, producing at last not more than 20 to 30 eggs: in general, it is most profitable to dispose of hens whilst they are yet saleable for that purpose, which is in the spring of the third year: nor do delicate white hens lay so many eggs in the cold season as the more hardy coloured varieties, though we believe that the white-feathered fowls are not of such extremely delicate nature as is popularly supposed, notwithstanding that most poultry-wives assert that they require more warmth and shelter, particularly by night. Hens *above the common size of the respective varieties to which they belong*, are by no means preferable either as layers or sitters.

The health of fowls is observable in the fresh and florid colour of the comb, brightness and dryness of the eyes, the nostrils also dry and free from any discharge, and the plumage of a firm, (not coarse,) close, and glossy texture. The indications of old age are, paleness of the comb and gills, dulness of colour, a sort of stiffness in the down and feather, length and size of the claws, and the scales upon the legs becoming thick, large, and prominent. Whilst under the natural course of moulting poultry are quite unfit for the table, as well as for breeding: it is the same also with respect to young fowls whilst shedding their feathers in early spring.

In buying fowls for the table, some people assert that the *colour of the legs* indicates, with tolerable certainty, the quality of the flesh,—an opinion sanctioned by that " good-living " man Soyer, who informs us that, for roasting, fowls with black legs are to be preferred, whilst those which have white or very light-coloured legs are the best for boiling. Yellow-legged fowls are often of a tender constitution, and always inferior in the quality of their flesh, which is of a loose, flabby texture, and ordinary flavour. It is generally held that white-feathered fowls have a great tendency to exhibit a yellowness of skin, fat, and flesh, by no means agreeable to the sight when placed on the table. Small-boned, well-proportioned poultry, greatly excel the large-boned, long-legged kind, in colour, fineness of flesh, and delicacy of flavour: for it is held good, that of all animals of the domestic kind, those which have the smallest, cleanest, finest bones, are in general the best proportioned, and are covered with the best and finest grained meat—besides being, in the opinion of good judges, the most inclined to feed, and fatted with the smallest proportionable quantity of food, to the greatest comparative weight and size. In the choice of full-sized fowls for feeding or fattening for the table, the short-legged and early hatched always deserve a preference: the green linnet is an excellent model of form for the domestic fowl, and certainly the true Dorking approaches the nearest to such model. Fowls reared for the table should be plump, deep, long, and capacious in body, with short legs, small-sized bones, very white, juicy, fine grained flesh, the fat and skin equally white, and of delicate flavour. As we have already seen, some varieties of fowls distinguish themselves as excellent sitters and rearers of young broods, whilst others are only remarkable for their abundant produce of eggs, scarcely ever desiring to incubate: it will, of course, be for the skilful and judicious breeder to avail himself advantageously of this variation in quality,—the latter supplying plenty of eggs, whilst the "sitters" furnish the means for hatching them.

Goose.—In selecting stock Geese of the common kind, those should be chosen that are long in the body and small in the bone. As a general indication, the loose abdominal pouch or bag, sagging down behind, may be taken to be a sign of age in the Goose; except in the Toulouse, which has it naturally very large, even from six months of age.

Duck.—In the choice of Ducks, the same rule as to shape should be observed, as we have recommended for Geese. Our old housewives had a notion that the variety of Ducks which have the bill bending upwards, lay a greater number of eggs than common; of which we can say nothing from observation, but Ducks well fed will rarely fail to have plenty of eggs. For their flesh, Soyer tells us, that the *least gaudy* plumage Ducks should be selected.

PRODUCE AND PROFIT.—Everybody knows the value of a fresh-laid egg, but in spite of the intro-

duction of the Cochin China breed, during the
winter months a supply of such a luxury cannot
be hoped for; and as eggs from the porous nature
of the shell emit or transpire, their watery parts
become desiccated, and in consequence decompose
—or, at the very best, lose a great part of their
substance and nutritive quality by keeping. But
it has been satisfactorily ascertained that they will
retain their moisture and goodness three or four
months, or more, if the pores of the shell *be closed*
and rendered impervious to the air by some unc-
tuous application; and being articles of such general consumption, it is a matter of importance to
housewives, to be able to preserve any surplus
summer stock to send to market in the cold season,
when of course their scarcity enhances greatly
their value: for this purpose those laid after harvest, from August to October, are generally put on
one side. Various plans are adopted, and find
their respective advocates,—some put them in
bran and salt, others insulate them in heaps of
corn, rye, sawdust, straw, ashes, &c., but none of
these have been found to answer well for a long
time, or on a large scale. Many years since the
following successful experiment for their preservation was made at Paris: a large number of eggs
were placed in a vessel, in which was some water
saturated with lime and a little salt; they were
kept in that state several years, and, being opened
in the month of January, were found in excellent

preservation, without a single failure. This account was forwarded to the author by an English lady, an experienced poultry rearer; and it has since been very extensively practised by the poultry-wives of our own country. Another method has lately been recommended,—to dip eggs in oil, and pack them in salt. At any rate they ought not to be deposited on their sides, which causes the yolk to adhere to the shell. A plan generally adopted, and which has been found to succeed better than almost any other, is to anoint the eggs carefully and gently with soft melted mutton suet, or pork lard, just sufficiently warm to rub on softly with the finger, and then to place them with the small end downwards, upright and wedged close together, strata upon strata, in an old fig drum, butter firkin, small cask, box, or any other vessel most convenient or readily obtained, covering the same over carefully when filled: they thus come into use, at the end of a considerable period of time, in a state almost equal to new-laid eggs for consumption, but ought not to be used for incubation, excepting in the case of the imported eggs of rare birds. When it is required to preserve eggs on an extensive scale, it will be found far less troublesome to place the eggs at once in some deep vessel, and then to pour over some melted butter (either salt or fresh) over the strata of eggs, in as cool a state as it can well be in order to run over; by this plan the butter would not be in any way injured by the contact with the eggs, (provided they were placed in clean,) and would therefore be equally useful for winter consumption: if, however, a cheap and economic mode of preservation be alone required, fresh rough mutton tallow, melted down and strained, may be conveniently substituted in place of the butter. In any case, the box or vessel containing the preserved eggs should be placed in a cool dry situation, until required for use or sale. It is said, but we have no knowledge on the subject, that pullets' eggs, which have not been fecundated by intercourse with the cock, will keep much longer when preserved than hens' eggs.

Feathers and down form no unimportant addition to the value of poultry produce, living or dead. Such as are intended for use should be plucked as soon as possible after the bird is dead, and before it is cold, otherwise they are defective in that elasticity which is their most valuable property, and are liable to decay. The bird should, besides, be in good health, and not moulting, for the feathers to be in perfection: and being plucked, and a sufficient number collected, the sooner they are dried the better, since they are else apt to heat and stick together. They may either be dried in a room by the sun, or in the oven, or they may be kiln-dried in bags or sacks; they then require to be well beaten or threshed with canes or small sticks, before being taken out of the bags, in order to detach any hard lumps or dirt from them, and when taken out passed through a sieve, and finally the stumps of the quills cut off with a pair of scissors.

The practice of plucking the *living fowl*, if interest must sanction such a custom, should be performed in a tender and careful manner. The ripe down only should be taken from each wing of the swan, goose, or duck, and four or five feathers. Lean geese furnish the greatest quantity of down and feathers, and of the best quality: to which also the goodness of their food, and the care bestowed, contribute in a considerable degree. Geese are sometimes stripped three times in the season, but in the whole affair we speak with entire ignorance of the practice. Strict precaution is necessary to house the stripped fowls, for a time sufficient to enable them to endure the air, and by all means to keep them from the water. M. Parmentier proposed to multiply the breed of white turkeys, and to employ for plumes the feathers found on the lateral part of the thighs of those fowls.

The following information, as to the practice in Lincolnshire and the Fen counties, was obtained some years since in answer to inquiries made on the spot:—" The geese are usually plucked five times a year, though some pluck them only three times, and others four; commencing at Lady-day, again at Midsummer, Lammas, Michaelmas, and

Martinmas. Goslings are not spared; early plucking, they say, tending to increase the succeeding feathers. The common mode of plucking live geese is considered barbarous; but it has prevailed perhaps ever since feather-beds came into general use. In answer to the charge of cruelty preferred against the 'fen slodgers,' the writer deems it an act of justice to state, that the owners are careful not to pull until the feathers are *ripe*, that is, not until they are just ready to fall; because, if forced from the skin before, which is known by the appearance of blood at the roots, they are of inferior value; those plucked after the geese are *dead*, are affirmed to be of still less worth. The larger feathers and quills are pulled twice a year only."

A writer has humanely remarked on the cruelty of plucking the living goose, proposing a remedy, which we should rejoice exceedingly to find practicable and effective. The remedy proposed, on the above authority, is as follows:—feathers are but of a year's growth, and in the moulting season they spontaneously fall off, and are supplied by a fresh fleece. When, therefore, the geese are in full feather, let the plumage be removed, close to the skin, by sharp scissors. The produce would not be much reduced in quantity, whilst the quality would be greatly improved, and an indemnification be experienced, in the uninjured health of the fowl, and the benefit obtained to the succeeding crop. After this operation shall have been performed, the down from the breast may be removed by the same means.

In speaking of the PROFITS of poultry-keeping, it would be easy to add to the statistical papers so often paraded in works on the subject, and at the same time to increase the mass of extravagant misapprehensions, which have been raised by the immense profits netted (or rather said to have been) upon the sale of stock or produce; and which to a great extent overlays the truth which falls to the bottom. Unfortunately (or fortunately) these fallacious data and consequent conclusions, seem a necessary attendant upon any violent setting-in of public opinion in any particular direction; and truly even this may be useful, for if it has the effect of directing public attention to a branch of rural economy hitherto almost entirely neglected, the object attained will be a good, and for once we may admit that, in this matter if not in others, *the end justifies the means.* That poultry may be reared not only without a loss, but even with some little profit, we readily admit; but without desiring to depreciate the somewhat arbitrary and fictitious value which the " Fancy " have placed upon certain kinds of fowls, (for we see no great objection to a Cochin China being sold at fifty guineas, any more than there is to a racer being sold for a thousand guineas,) but assuredly it is not by giving such prices, or even ten or twenty guineas a pair, for stock, that the amateur can hope to realize a handsome profit; and not less absurd would it be for him to expect to make poultry-keeping "pay," by erecting *cottages ornées* for his fowls, aërial palaces for his pigeons, or ornamental aviaries for his pheasants. The statements put forth (perhaps in perfect good faith by the parties themselves) in various works, as far as we have seen, have been based upon such fallacious premises and data, that it would be worse than useless to reproduce them in our pages, as in any way correctly exhibiting the assumed profitable nature of amateur poultry-keeping: we however gladly give insertion to the following carefully recorded account of the relative *cost* and *profit* of a small stock of poultry, upon an experimental and model farm at Ankerwycke House, Buckinghamshire, the seat of George S. Harcourt, Esq., by whom we have been most obligingly favoured with it; and it has this merit at least, that it is not introduced to serve any particular view, but presents a reliable summary, deduced from actual facts,—in fact, a concise balance-sheet, based upon no assumed or supposititious transactions, all the stock and produce having been actually sold or disposed of in the way described,—none being consumed.

FROM MICHAELMAS 1851, TO MICHAELMAS 1852.

Dr.	£	s.	d.	Cr.	£	s.	d.
14 fowls bought for stock birds	1	8	0	To 1326 eggs sold at 1s. per dozen	5	10	6
Quarter of Oats consumed for feed	1	4	0	To 14 fowls sold at 2s.	1	8	0
Quarter of Barley consumed for feed	1	8	0	To value of 9 fowls added to Stock		18	0
	4	0	0				
Profit to balance	3	16	0				
Total	7	16	0	Total	7	16	0

The fowls were of the so-called " Improved Dorking " or Sussex breed; and under the management of a careful and experienced man, but with only natural feeding, no extraordinary means being resorted to, in order to increase their productiveness, which in all probability might have been considerably increased by high or over-feeding.

It must be evident that wherever the main source to be relied on for poultry feed is *purchased* grain, and the entire cost of houses, loss of time, or services of attendants, is taken into account, it would indeed be vain to expect the pursuit could be profitable: but in every farm the offals or refuse of the barn and stable will maintain a certain number of poultry: these, as they are fed upon what must otherwise be lost, can only be regarded as mere save-alls, and as they cost the farmer scarcely anything, almost all that he gets is pure gain; whilst the occupation of rearing them affords at once a healthful amusement and social enjoyment for his wife or children. Again, as the poorest family can often maintain a cat or a dog in their humble households, without any apparent expense, so even the poorest cottagers or occupiers of lands or allotments can generally rear a few poultry at a perfectly inappreciable cost,—the little offals of their own tables supply them with a great part of their food, and they will find the rest in the neighbouring fields and lanes, without doing any sensible damage to anybody.

Upon equally fallacious data innumerable statements have been published, instituting comparisons of the *relative profitableness* of each variety; although such have been often made under special and even totally dissimilar circumstances, with perhaps not a passing consideration as to the relative situation, condition, feeding, &c., to which the fowls have been subjected.

In reference to the comparative profitableness of different stock, our ample notices of the qualities of almost every variety, in another division of our subject, almost renders any statistical comparisons unnecessary.

That the Cochin China fowl surpasses every other as a winter layer, there cannot be a reasonable doubt. The following statement of the produce of forty Dorkings and twenty Cochins upon the model farm of George S. Harcourt, Esq., is valuable for the reasons previously given : the return is from a register, most carefully and accurately kept, and extending from the 1st September, 1853, to March 31st, 1854.

	Twenty Cochins laid	Eggs.
1853.	In the month of September . . .	56
— —	October	68
— —	November . . .	50
— —	December . . .	141
		315
1854.	From January 1st to March 31st . .	532
	Total produce in 7 months	847

	Forty Dorkings laid	
1854.	From January 7th to March 31st . .	521

N. B. They did not lay from September to December, 1853.

By the above figures it will at once be apparent, that during the whole period the *proportionate* produce of the Cochin China was more than three times that of the Dorkings; whilst, in the most favourable light, if we confine the comparison to the months only during which the Dorkings laid, the *proportionate* produce of the Cochins will even then be in the ratio of 2 to 1 of the Dorkings. It must not, however, be concluded that the Cochins are *all* profit: they have one drawback, namely, the greater quantity of food which they consume above most other fowls,—for, when young especially, they are great feeders, and undoubtedly consume more food than other varieties, (the Malay perhaps excepted,) for it is absurd to deny the fact : eggs have been rightly described to be " the superfluity of the animal's nutrition—the profitable balance, in fact, of the stock of provision it consumes ; " and it is useless then to expect any secretion of surplus nutrition, in the shape of abundant produce, unless precedent abundant feeding is allotted to the fowl ; and in addition, their very nature—their peculiarly quiet and domesticated habits—induce them to remain within very circumscribed limits, rather than to search for those minute insects and worms, and stray grains, from which other fowls derive no inconsiderable portion of their sustenance.

The Game variety, though surpassing in point of beauty and richness of feathering, are far too pugnacious to be *profitable ;* as, even when kept by themselves, their frequent and deadly conflicts entail much trouble and loss in the yards ; at times whole broods, scarcely feathered, will be found stone-blind from fighting, to the very smallest individuals ; the rival couples moping in corners, and renewing their battles on obtaining the first ray of light. On account, therefore, of the extreme difficulty of rearing the chickens, from their natural pugnacity of disposition, which shows itself at the earliest possible period, breeders are deterred from rearing them : and as this disposition, to a certain degree, prevails in the half-bred, it prevents crossing with the Game cock, otherwise a great improvement.

Without offering our own opinion, we may at least mention that of many experienced breeders, that, *in proportion to the quantity of food consumed*, the Dorkings will, on the whole, be found more profitable both as regards the eggs and their value as table fowls: upon the latter point it is acknowledged they have no superiors; and they moreover fatten earlier than most others, or at least they appear to be always so well covered with meat in proportion to their size, as to require less feeding to render them readily marketable.

The Spanish is confessedly among the most prolific layers, and of eggs of extraordinary size; but as those commodities are sold numerically, generally without reference to size, and certainly quite irrespective of superiority in regard to weight, it must be obvious that large eggs are not *relatively* quite so profitable as small,—as they are always less in number, and require more surplus feed for their manufacture.

Polands, though most beautiful and ornamental fowls, and withal good layers and not great eaters, cannot be said to be advantageous *as mere farm-stock birds*, on account of their being more tender and difficult to rear,—unless indeed upon fine dry uplands, with an extensive range, when they flourish with but little attention or care being required.

The Hamburgh (or every-day layers) may be characterized as most productive egg-manufacturers—excellent table fowls—and hardy in most situations; to which may be added, as regards the spangled varieties only, that they are fair brooders.

The Bantam, though a delicate viand for the table, is perhaps the furthest from being a *profitable* kind of poultry,—the eggs being almost unsaleable at market from their extremely small size.

The propriety of encouraging the rearing of Pigeons in an economical point of view, as a source of *profit*, has long been the subject of much dispute. M. Duhamel, the apologist of these beautiful favourites, avers, that pigeons do not feed upon green corn, that their bills have not sufficient power to dig for seeds in the earth, and that they only pick up scattered grain, which would else be wasted, or become the prey of other birds: from the season of the corn appearing, he says, pigeons subsist upon the seeds of weeds, the multiplication of which they must, in consequence, greatly prevent. No one, however, in the least acquainted with country affairs, but is fully aware of the immense damage done to the crops of corn, beans, pease, and tares, that is to say, the grand articles of human subsistence, by pigeons. Our best practical agricultural writers may be consulted on this head, but a sufficient proof of the fact is the reduction of dove-cotes throughout all counties where agriculture is best known, valued, and practised. To the value of the agricultural produce of the country, must also be added the injury committed by pigeons in seed-time, by picking up grains of seed, wherever they alight, and the corn trodden under and beaten out by their wings before harvest, not to forget the real damage they do to buildings, by pecking the mortar from between the bricks and tiles; whilst in the garden they may be otherwise destructive. On a general view of the subject therefore, it appears that the dove-cote system cannot fail of being injurious to the agriculturalist, without any commensurate or compensating advantages; for it must often unavoidably happen, that great flocks will be maintained at the expense of persons having no property in them. But as, certainly, neither the public nor individuals will consent to be deprived of the enjoyment of this ancient luxury, the fairest mode appears to be, the regular feeding of pigeons by their proprietors; which so attaches them to home, that there is often a necessity of driving them out for exercise. This plan should, of course, be more punctually observed in seed-time, and towards the approach of the corn crops to maturity: but to prevent effectually the risk of damage from pigeons, which will necessarily be incurred, the farmer must practise that vigilance in which generally he is too defective. One word, however, may be said in favour of the *columbine* family: if reared in a poultry yard *in towns*, they do no damage to any, afford an agreeable varia- tion and addition to our table-dishes, and require scarcely any attention worth mentioning, foraging all the food they require from the scattered meals of the fowls. If, however, the original cost of the birds, and the *value* of the food they consume, be estimated, no one who has had the slightest ex- perience of them can for a moment pretend that Pigeons can be kept and reared *profitably*.

Turkeys are the most tender and difficult to rear of any of our domestic fowls; but with due care and attention—which, rightly considered, in all things give the least trouble—they may be produced and multiplied with little or no loss; and the same may be averred with all truth of the rest of our domestic fowls, and animals in general, the losses and vexations annually deplored arising almost entirely from ignorance and carelessness united hand in hand. Turkeys, under a judicious system, may be rendered an object of a certain degree of consequence to the farmer. They share with geese in gleaning the corn-fields, or " shack- ing," as it is termed; they will also forage over woods and commons in the autumnal season, after which they are put up to be completely fattened. The Black Turkey, as we have elsewhere observed, is generally most esteemed, on account of the su-

perior whiteness of the skin and flesh, the fineness
of the meat on the breast, and being of a sweeter
and higher flavour; it is also thought that the
males attain a greater size, and that the females
are better breeders, than the lighter coloured
fowls; at any rate many of the most intelligent of
our poultry-wives refuse to rear any but the
Black, being persuaded in their own minds that
they are the most profitable. The age of Turkeys
is indicated by the scales on the legs: these when
the bird is young are smooth and soft; when old,
they are rough, hard, and rigid.

Guinea-fowls, from their wild, untractable,
and rambling dispositions, have been placed under
the ban and excommunication of almost the whole
poultry world; though we believe, if properly
managed, they would prove a most useful and
profitable description of poultry, from the number
of eggs they lay, and the ready sale the birds themselves command for the table. If allowed to roam,
moreover, they will find half their food—indeed,
whether allowed to do so or not, is of little moment, they will by some means or other forage up
half what they require.

Of both Geese and Ducks it may be said that
they are, generally speaking, reared with very
little trouble, and a fair amount of profit; they
are alike valuable to the small farmer, the cot-
tager, or the allotment holder, or indeed to any
one living within range of waste grounds or com-

monage, where they may be maintained at little cost either as regards food or trouble of management. The Goose has *par excellence* a claim to be included among domestic poultry—which it is in every sense, whether as regards habits, utility, or economic value. Nobody requiring Geese would dream of breeding them for their beauty or their plumage—size and weight being the invariable requisites sought for, and these can be always attained under proper management cheaply and readily,—as they shift for themselves almost from the very shell.

Ducks, from their aquatic nature, have the merit of thriving exactly in those places where all other kinds of poultry will not, in damp, swampy districts. In almost all rural sites or marshy places —in fact, every where in the proximity of a pond, dyke, or other run of water, they may be kept with profit; for although they have been charged (and justly) with being most enormous eaters and inordinate gluttons, they are satisfied with, and indeed greedily devour, all kinds of offal and garbage which other fowls would turn up their bills, if not their noses at. They will also clear away slugs and snails from gardens, whereby, however, it must also be admitted, they will do much damage to the young greens unless looked after carefully: ponds or pieces of water covered with weeds can easily be cleared by placing a few ducks upon it, for which purpose they are useful.

CHAPTER VIII.

POULTRY HOUSES AND YARDS — BREEDING — FATTENING —
INCUBATION — HATCHING — REARING — FEEDING, AND GE-
NERAL MANAGEMENT OF POULTRY.

IT has been already observed, that the warmest
and dryest soils are best adapted to the breeding
and rearing of gallinaceous fowls, more particularly chickens: thence the greatest success, attended with the least trouble, may be expected on
such; and far greater precaution and expense will
be required on those of an opposite description.
Of these last, the wet and boggy are the most injurious, since, however ill affected fowls are by
cold, they endure it still better than moisture;
whence they are found to succeed well upon dry
land, even in the severe climates of the north.
The warmer climates are far more favourable than
ours for the purpose of raising poultry, and the
same rule necessarily holds with respect to our
own country, where it has been found that upon
dry and shingly land, like the sea-beach, and
where scarcely any care was taken of the breeding
stock, or shelter afforded to them, yet they multiplied in a most extraordinary degree, and were
preserved in a constant state of good health. Upon

a boggy or clayey soil, under such circumstances, they would have died like rotten sheep. In short, land proper for sheep is generally also adapted to the successful keeping of poultry.

But as the rearing of them is necessary, upon soils and in situations of every description, it will be most to the purpose to point out those precautions which must be recurred to, in order to insure success upon the least favourable. Of such, then, artificial or made ground cannot be dispensed with for a poultry-yard, where rearing is made an object upon any considerable scale; since upon damp and boggy soils, not only will the greater part of the broods be annually subject to disease and mortality, but the cocks and hens themselves will be frequently affected, to the great impediment of the business of the breeding season. Where it is not held worth while to make any extraordinary accommodations for poultry, and the risk taken, enough may yet be preserved for family convenience and to repay the trifling expense. But no considerable stock can be kept, far less any profit made upon it, upon an unfavourable soil, independently of attention to needful local conveniences.

As there are perhaps no two country or farm houses laid out or constructed exactly alike, of course a person entering upon them, and intending to rear poultry, must, to a great extent, take things as he finds them, and avail himself of the existing buildings and conveniences as far as possible : it seems therefore useless to follow the example of

our predecessors by laying down arbitrary rules
for the construction of poultry-houses, which,
under local or other circumstances, are almost
always impracticable. Neither is it our intention
to give a series of plans and diagrams of the regal
and princely poultry-houses of the Queen and her
nobles, which, although they make an imposing
appearance in works of this nature, would seem
calculated, by their extent and costliness, to display their owners' taste and extravagance in laying
out their money and the yards at the same time,—
rather than to effect any object of practical good
or utility, by serving as a model or design for
poultry-keepers and amateurs. Where, however,
poultry are introduced and reared for the first
time on the spot, it is obviously necessary that
some instructions and plans should be given to
the inexperienced in order to enable them to build
the houses, lay out the yard, and provide the requisite conveniences for the reception of the stock
it is proposed to rear. In this view therefore we
have annexed some ground plans of such, which,
though having no pretensions to be considered
very elaborate or elegant, have at least the merit
of being simple in design and construction, cheap
in cost, and adapted to insure, under proper management, the well-doing of the stock : the references will, we think, render it perfectly intelligible
to our readers—at least much more so than any
mere description could.

PLAN OF TURKEY, GOOSE, AND DUCK HOUSE.

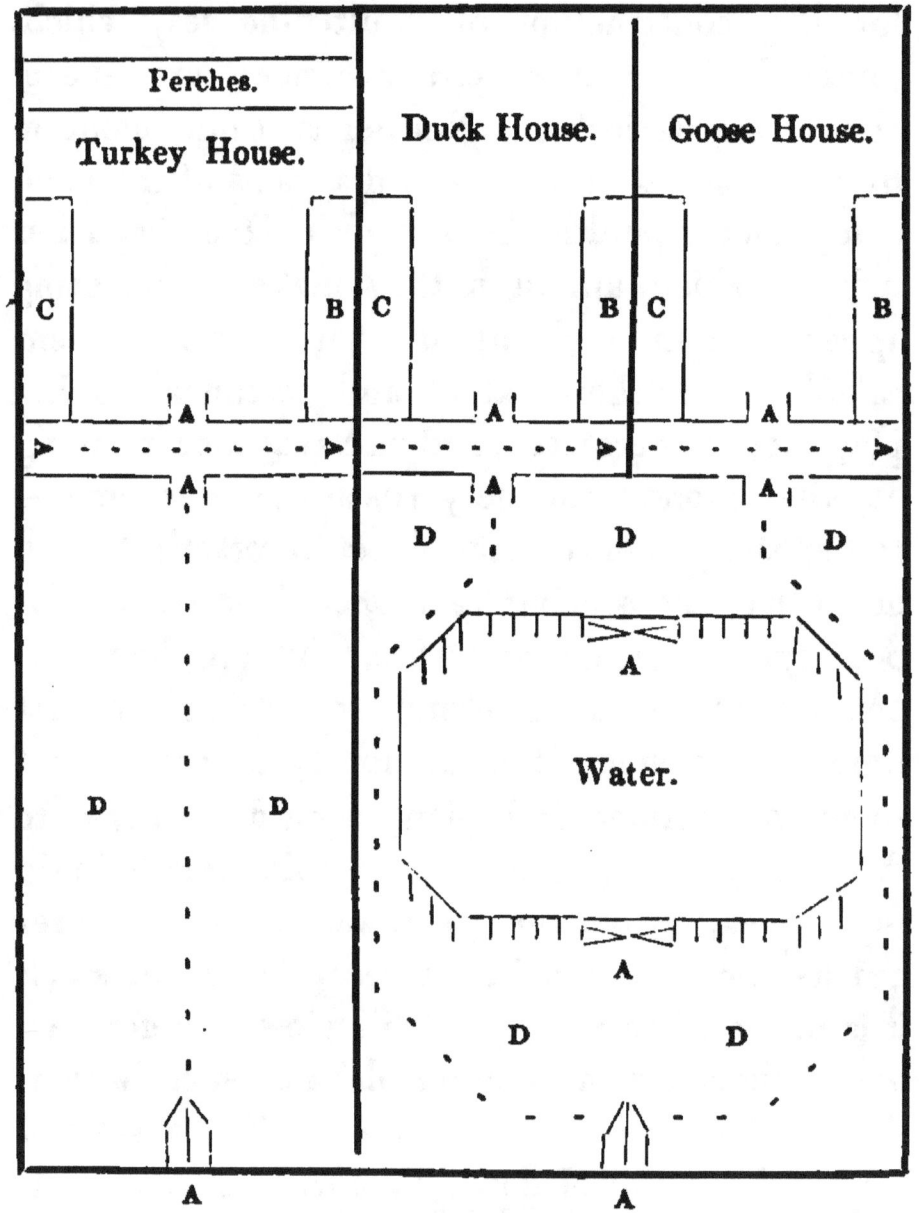

REFERENCES.

A—Doors and wicket gates.
B—Laying boxes or nests.
C—Hatching boxes or nests.
D—Grass plots or walks.

The *dotted* lines - - - - - indicate pathways unenclosed, but crossed with wicket gates.

PLAN OF FOWL HOUSE.

Perch.	Perch.	Perch.	Perch.
Fowl House.	Fowl House.	Fowl House.	
C B	C B	C B	Fattening House.
A	A	A	A
D	D	D	D

Dove Cotes.

Dove Cotes.

Dove Cotes.

Dove Cotes.

A   A   A   A

REFERENCES.

A—Doors and wicket gates.
B—Laying boxes or nests.
C—Hatching boxes or nests.
D—Grass plots or walks.

The *dotted* lines - - - - - indicate pathways unenclosed, but crossed with wicket gates.

Whether or not the poultry be suffered to range at large, and particularly to take the benefit of the farm-yard, a separate and well-fenced yard or court must be pitched upon. The foundation should be laid with chalk, or bricklayers' rubbish, the surface to consist of sandy gravel, considerable plots of it being sown with common trefoil, or wild clover, with a mixture of burnet, spurry, or star-grass, which last two species are particularly salubrious to poultry. The surface must be so sloped and drained as to avoid all stagnant moisture, most destructive to young chickens. The fences must be lofty, well secured at the bottom, with wire or twine netting, or laths, that the smallest chicken cannot find a passage through, and the whole yard perfectly sheltered, from the north-west to the south-east: we prefer the galvanized wire netting as fencing, not only on account of its not being affected by the wet, but because an objection exists to the twine, in consequence of young fowls and pigeons getting their heads through the elastic meshes, thereby often hanging themselves, which from the unyielding nature of the wire they are unable to do. Various beds or heaps of sifted ashes, or very dry sand, should be always ready, in which the fowls may exercise that propensity, so delightful and salutary to them, of rolling or bathing themselves. A butter-firkin, sawn in halves, each half sunk flush with the surface of the ground, and placed in different parts of the yard, serves admirably to form receptacles for the "dusting bath:" the best ashes are those of wood or coal burnt very fine; or else, when these cannot be obtained, dry sand, burnt lime, or ground oyster shells, will be a convenient substitute. In order to render this necessary convenience always available, it is requisite that a small low lean-to shed should be constructed over it to keep the wet from it,—it may be of the simplest kind, a few boards placed over four stakes, about 3 feet from the ground, in a corner of the yard, will serve the purpose, and will also provide shelter for the fowls in rainy weather, as they will otherwise often stand out in the wet rather than take refuge in their night houses during the day. The bath is effectual in cleansing their feathers and skin from vermin and impurities, promotes the cuticular excretion, and is materially instrumental in preserving their health.

The poultry-houses within the court, if there be a choice, should have a southern aspect, at any rate should be well defended from cold winds and the blowing in of rain or sleet. If the number of the stock be considerable, the houses had far better be small and separated, both for the sake of health and safety; and especially they should be absolutely impenetrable to vermin of every description. Should these houses abut upon a stable, brew-house, or any conductor of warmth, it will be so much the more comfortable and salutary to the poultry: but heating by stoves or flues we consider generally unhealthy.

The form and conveniences of the poultry-houses are these—the bottom or floor should consist of well-rammed chalk or earth, similar to the court-yard, that its surface, being smooth, may present no impediment to being swept perfectly clean. For health's sake, the roof should be lofty; the perches will be then more out of the reach of vermin, should any accidentally break in: and there should be only one long and level range of perches, extending longitudinally through the building; because when these are placed one above another, the fowls dung upon each other: convenient steps driven into the walls, will render easy the ascent of the poultry to their perches; but care must be taken that the mistake be not made of placing these steps immediately one over the other, but in such wise, that they can jump from one to the other.

Of the dimensions of the yard or house we will only say that they cannot well be too large, (especially the former,) and therefore parties must be guided by the conveniences and extent of ground at their disposal, as well as the description of stock required to be accommodated: let us, however, expressly guard against their being built *too low*—which is most injurious to the health of the fowls; the roof of some water-tight material, tile or slate, (the latter the best, as being warmer

in winter and cooler in summer,) and fixed sloping
down to the front, at a minimum height from the
ground of 7½ feet at the back, and 7 feet in the
front, whence a common zinc or other pipe should
be placed to convey the rain-water from the eaves;
the roof may be lighted by a glass tile or slate, if
desired: some writers recommend thatch for the
covering of the roofs, but apart from the danger
occasioned to it from fire or lightning, it is much
to be condemned as affording refuge for lice and
various insects. At each end of the fowl-house, a
lathed-over window, about one foot square, should
be cut, about three inches from the roof, and similar apertures (also latticed over) formed in the
partitions of each compartment into which the
house is divided, in order to secure perfect ventilation when the doors are shut: the windows at
the *outer ends* of the house should have slip-
shutters to close in cold or damp weather,—the
inner or partition windows, which of course only
communicate inside, and circulate the internal
air, need no shutters. The size of the perches
must be accommodated as nearly as possible to the
description of fowls,—the best form is an oval (the
flattest side upwards) or octangular wooden pole;
for the ordinary varieties it may measure about 9
inches round, but for the Cochin China, or any
very large fowls, they should not be less than 12
or 13 inches in circumference, as, if smaller, the
long toes and claws of those fowls prevent their

getting a firm footing or grasp on the perches: a very simple mode of forming perches for fowls, is by sawing a long fir pole in halves, and turning the rough side (or that with the bark on) upwards. For the Cochins and other *large heavy* poultry it is advisable not to place the perches anything like so high as for others—certainly not more than 2 or 3 feet from the ground: we do not recommend this for the reason usually given, namely, that from resting on the perches they become crooked-breasted, which is manifestly absurd, because whether the perches were high or low, if they rested with their breasts upon the poles, the consequences must in either case be precisely the same; fowls, moreover, of considerable growth, will many of them become crooked-breasted when perching at large; and much depends upon form in this case, since we find that aged fowls, which have perched all their lives, have often the best shape, and the breast-bone perfectly straight: but our objection to high perches is, that large heavy fowls are deficient in the powers of flight, and therefore have considerable difficulty in ascending, often falling heavily to the ground two or three times before they gain the perch; and again in descending, being unable to sustain themselves or break their fall. The front doors or entrances to the house itself, should of course be of man's height, having a slip-to wicket in the bottom, about 15 inches high by 10 broad, in order to admit of the ingress and egress of the fowl when inclined, and at such times when the door would be too much to open. The floors should slope down to the doors in front, in order to get rid of the refuse water by means of drains immediately under the doors, running along the pathway.

Boxes, of which every carpenter knows the form, are to be arranged round the walls, on each side, in sufficient number. The board, or step at the entrance, to be of sufficient height to prevent the eggs from rolling out. Provision of a few railed doors may be made, for occasional use, to be hung before the entrance, in order to prevent other hens from intruding to lay their eggs in it. The tops of the laying and hatching boxes should be covered over with a shelf or board, sloping down from the wall to the front, in order to prevent the birds roosting or dunging upon it; and at the same time capable of being removed, when requisite to renew or clean out the nests or boxes themselves. Some breeders construct the boxes of slate instead of wood, which they say is kept clean easier. A very simple plan, may be contrived by placing some of the common large brown earthen dishes, of an oblong square, about 4 inches deep, in the lockers at the sides; as they could be shifted and cleaned with greater facility than any other. The common deep square boxes, uncovered at top, are extremely improper, because that form obliges the hen to jump down upon her eggs; whereas for safety she should in a manner walk in upon them. The same objection lies against hampers, with the additional one of the wicker-work admitting the cold. Some breeders prefer to have all their nests upon the ground, on account of the danger of chickens falling from those which are placed above.

Although we are aware that it is usual to give numerous specific directions for the formation of houses for particular breeds,—which only confuses, —if the hints we have already given be observed, we venture to say that almost any kind of poultry will be found to do well in houses of a similar kind, as nothing is more absurd than to suppose that each individual variety of fowl requires peculiar houses, or a specific treatment. Although it is difficult to say in how limited a range poultry may be reared, still all table fowls require a much larger and more extensive range than the Cochin. Where Polands are required to be reared, the site for the yard should be chosen on the sunny side of some nice dry uplands. Cochin Chinas seem to do well wherever any other kind of fowl can; all that requires to be noted is, that if allowed to nest on the ground instead of perching, they will be very comfortable. But after all, the grand secret of *all* fowls' well-doing may be summed up in these conditions—cleanliness, dryness, and warmth of houses, a comfortable range, and good (not exciting) food; and a little careful attention to these points will effect more than the perusal of volumes of rules of scientific experimentalists, or the study of learned disquisitions upon nice distinctions between tiles or slates for the roofs of the poultry-houses, or the preference to be given to high or low perches.

Pea-fowls are so seldom kept in poultry-yards, that it seems useless to speak of the houses they require : if desired, however, they may be accommodated like turkeys. They cannot be kept to their sheds, even in the winter sometimes ; and as for the summer, they always (if possible) roost abroad upon the highest trees they can find, or, in default of any at hand, upon some neighbouring house-tops.

Turkeys may be kept either in a separate house, or their boxes for laying or sitting may be placed upon the ground of the fowl-houses; which last method, perhaps, is objectionable, since turkeys and common fowls might not roost quietly together. Turkeys getting into a field of corn, will do much mischief by beating it down. Turkeys in the neighbourhood of large woods, if not watched and prevented, will eagerly stroll thither without any desire to return : and they prefer roosting abroad upon high trees in the summer season, if that could be permitted with a view to their security or safe keeping. Turkeys, from their size, require much stouter perches than other fowls. Poultry of all kinds are sometimes associated in a common house, the cocks and hens aloft, and the ducks, geese, and turkeys upon the ground-floor. Or, upon an extensive scale, all the domestic poultry may be contained within the enclosure, a circular form being perhaps the most comprehensive and advantageous; including a piece of water, with laying-houses upon its banks, for the aquatic fowls, and dove-cotes for breeding pigeons.

Under a regular system, it would be preferable to separate the aquatic from the other kinds of poultry; the former to have their houses ranged along the banks of a piece of water, with a fence, and sufficiently spacious grass plots or walks in front: the access to the water to be by doors so arranged that they may be closed at will. Should the water be of considerable extent, of course a small boat would be necessary, and it might also be employed as a means of conducing to the pleasure and amusement of angling. After severe frosts covering the water, the ducks must not be let out before the ice has quite disappeared; else they are apt to be lost by diving beneath floating blocks. They are also not safe (especially the young ones) in waters stocked with eels; and as to pike, every one initiated in country concerns, knows very well, that those voracious fish will make free alike with ducklings, goslings, and cygnets: few duck-ponds, however, have pike in them, though many have eels; and but few breeders, we fancy, would trust their stock in waters where there are pike.

Swans wandering by night, in search of water-cresses chiefly, are always in danger from the different vermin which prey upon poultry and game —weasels, stoats, pole-cats, &c. And swans thus destroyed exhibit no wounds or marks upon the body, but upon the head and neck, where, on inspection, the wounds are discovered, though the feathers are unruffled. The wounds given by the sharp and long teeth of the vermin, appear scarcely the size of a pin's head, but are generally above half an inch deep. Geese and turkeys are also liable to be destroyed by these marauders.

Where geese and ducks are also kept, it will, of course, be requisite to provide houses and yards for them also: a glance at the plans we have already given will show how easily and conveniently these may be arranged either by the side of the fowl-houses, or back to back of them, if circumstances called for such arrangement. To rear geese successfully it is by no means necessary, *absolutely*, that a stream or even a large pond should be provided in the yard, as otherwise we should not see so numerous and flourishing flocks upon our waste commons, where no such additional advantages are often found: a clean, dry, well-littered house is all that is required, a small reservoir of water supplying the place of a pond where the latter cannot be obtained, as after all we must admit such a piece of water forms a valuable acquisition. It will be necessary to destroy all the hemlock or deadly nightshade within range of young geese, many of which drop off annually from eating it. We believe also that both the young and old have been occasionally killed by swallowing slips of yew. In regard to the first range for young goslings, a convenient field, containing a little water, is to be preferred to an extensive common, over which the gulls (or goslings) are dragged by the goose until they become so cramped or tired, that, squatting down and remaining behind at evening, they are lost.

In each division of the goose-house not more than eight or ten geese ought to be placed, as when more crowded the old will generally beat and tyrannize over the goslings. The nest is made upon the floor, of litter of straw, placed in the circular form of a nest. The floor may be made like that of the fowls, and strewed over with sand or *fine* gravel, the former being the better. The habitation for the ducks need not differ from that of the geese, except that as the duck delights only in water, the floor may be more often sluiced down to cleanse it, and therefore it will be better made of brick pavements, properly laid and inclined so as to drain off the water from the surface: it is not necessary, but it can do no harm, to strew down a little straw upon the clean floor; and the same material will serve for the nests. In the fens and also in Buckinghamshire many families used to rear large broods of early ducklings by the hand in their own cottages, a plan, however, we do not recommend.

No yards are complete without feeding-houses, for fattening poultry, warm and airy, with earth floors, as already described, well raised, and capacious enough to accommodate twenty or thirty fowls. The floor may be slightly littered down, the litter often changed, and the greatest cleanliness observed. Sandy gravel should be placed in several different layers, and often changed. A feeding yard well gravelled, and sown with grasses, is likewise a valuable adjunct to the house; being left open all day, for the fowls to retire in at pleasure.

In isolated parts of rural districts, all the above arrangements, the best concerted plan, and the most valuable stock, will little avail the proprietor, without a certain precaution,—a defence against midnight thieves. Not merely a lock, or a bar, or a mere trifling apology for security, but an ample safeguard: it is no unusual thing to be robbed, in a single night, of the greater part of a most valuable stock of poultry, the breed and excellence of which it may take several seasons to recover. In the first place, these small buildings should be made substantial; for, locks being good, thieves make their way by wrenching open an angle of the building. In addition to substantial locks and hinges, bells hung upon the inside of the doors, or upon any part liable to be shaken, are good precautions, since the noise may deter the thieves, even if it fail to alarm the family. But the most certain security is that kind of vermin cur generally kept by country labourers. Several such in the poultry court, being equally useful against robbers and vermin. Nothing can charm and quiet the tongues of real good *latratores*, or barkers, and particularly when several are together.

A plan like the above will obviously require the exclusive services of one or more attendants, according to its extent. Poultry may be well attended by an aged labourer, with the occasional assistance of his wife; and many a meritorious couple could thus be made easy and comfortable.

In regard to a dove-cote, or pigeon loft, the situation or extent must necessarily depend on convenience; one general rule, however, must be invariably observed,—that every pair of pigeons have two holes, or rooms, to nest in. Without this indispensable convenience there will be constant confusion, breaking of eggs, and destruction of the young. Pigeons do well near dwellings, stables, bake-houses, brew-houses, or such offices: or their proper place is in the poultry-court. A dove-cote is a good object, situate upon an island, in the centre of a piece of water: and the site may be rendered extremely beautiful and picturesque by planting, and a little simple ornamental and useful building. Where pigeons are kept in a room, some persons prefer making their nests upon the floor, to escape the danger of the young falling out; but in all probability guarding against the risk will incur that of rats and other vermin.

The front of the pigeon-room, or cote, should have a south-west aspect; and if a room be selected for the purpose, it is usual to break a hole in the roof of the building for the passage of the pigeons, which can be closed at convenience. A platform is laid by the carpenter at the entrance, for the pigeons to alight and perch upon, with some kind of defence against strange cats. The platform should be painted white, and renewed as the paint wears off, white being a favourite colour with pigeons, and also most conspicuous as a mark to enable them to find their home. The boxes also should be so coloured, and renewed as necessary, for which purpose lime and water will be sufficient.

Cleanliness is one of the first and most important considerations; the want of it in a dove-cote will soon render the place a nuisance not to be approached, and the birds, both young and old, will be so covered with vermin, and besmeared with their own excrement, that they can enjoy no health or comfort, and mortality is often so induced. Ours are cleaned daily; *thoroughly* once a week, a tub standing at hand for the reception of the dung, the floor covered with sifted gravel and sand, often renewed. Pigeons are exceedingly fond of water, and, having a prescience of rain, will wait its coming until late in the evening, upon the house-top, spreading their wings to receive the refreshing shower. When they are confined in a room, they should be allowed a wide pan of water, to be often renewed as a bath, which cools, and assists them to keep clear of vermin.

The common barrel dove-cote, which needs no description, is adapted to every situation in which it is desirable to keep pigeons. In the room or loft, the shelves should be placed sufficiently high for security against vermin, a small ladder being a necessary appendage. The usual breadth of the shelves is about 20 inches, with the allowance of 18 between each. Partitions may be fixed at the distance of about three feet, making a blind, by a board nailed against the front of each partition, that the pigeons may sit in privacy, and not be liable to be disturbed. Some prefer breeding-holes, entirely open in front, for the greater convenience in cleaning the nests; but it is from such that the squabs are likely to fall; thence a step of sufficient height is preferable. To make a nest, a straw basket or (which is better) earthen pan, may be procured, proportioned to the size of the pigeon being bred. Thus a pan three inches high, eight inches over the top, and sloping to the bottom like a basin, will be of sufficient size for a Tumbler, and one of double those dimensions for a large Runt. A brick should always be placed in contiguity to the pan, to enable the cock and hen to alight with greater safety upon the eggs. The pigeon trap on the house-top, is the well-known contrivance of those rascals who lie in wait to entrap the property of others; and who, abounding in almost all town districts, by their thievish activity, deprive the Fancier of the pleasure and amusement of flying his pigeons. A trap of another description, and for a very different purpose, is sometimes used; it is an area, on the outside of a building, for the purpose of confining in the air valuable breeds of pigeons which cannot be trusted to flight. Some are erected 20 yards long and 10 yards wide, with shelves on every side for the perching of the pigeons; thus they are constantly exercised in the air, retiring at their pleasure to the room or loft within. One of the simplest kinds of traps in use is a square open or lathed box, having a door on hinges letting down in front, a string being attached to it and passed into the loft, so that the owner can pull it up as soon as the captive pigeon has ventured within the precincts: this trap is usually affixed either in front of the loft, or at an aperture in the roof. We are thus particular in our description, not with the view of facilitating the capture of stray pigeons, but one is absolutely necessary where it is intended to fly birds, as the only ready means of egress and ingress from and to their own lofts.

BREEDING. *Fowls.*—The first step necessary

is most obviously to match the females to the males : the number of hens to be allowed to one cock, for breeding purposes, should be from four to six—the latter being the extreme number, with a view of making the utmost advantage. Ten and even twelve hens have been formerly allowed to one cock, but the produce of eggs and chickens under such an arrangement will seldom equal that to be obtained from the smaller number of hens. Spring is the best season to commence breeding with poultry; and, in truth, it scarcely matters how early, presupposing the best food, accommodation, and attendance, under which, hens may be permitted to sit in January. When the original treatise first appeared, many years ago, the poultry-wives of France made themselves very merry at the practice recommended of restricting the cock to so few as six hens,—their own allowance being from twenty to twenty-five hens; but we can assure the reader the rule was only adopted as the result of actual and long experience,—and since approved and carried out by many of our best poultry rearers. On the other hand, it may be well to guard against a cock being too much restricted, as if put with only one or two hens, the poor things get most terribly teazed, their plumage pulled to pieces, and themselves often seriously injured. To ensure good strong chicks, (and especially with all fine or high-bred birds,) a cock not less than two years old must be selected to breed from : the third year is the proper age for breeding, as almost all varieties of fowls are then in their prime and full vigour.

Nothing is more necessary towards success in obtaining plenty of eggs, than a good attendance of efficient cocks, especially in the cold season. If any of the cocks prove indolent, or at all inactive, the best plan is at once to withdraw the useless one to a separate ·walk, and substitute another which is known and familiar with the hens, since a stranger will not always be received.

The conduct of the cock towards his hens should be early and constantly attended to, as he will sometimes conceive an antipathy to a particular individual ; should this continue, the obnoxious one should be removed, since nothing but misery can ensue to the persecuted bird. Such a hen being removed, and replaced by a stranger, care should be taken, for the first week or two, that she be not worried or injured by the other hens. A place of refuge should be provided for hens or chicks in this unfortunate predicament. Whilst the young feathers are growing after moulting, poultry are extremely apt to peck and wound each other, retarding their recovery.

The change of a cock, from death or accident, is always attended with interruption and delay ; further, a new cock may prove dull and inactive from the change, however good in nature. This frequently happens with cocks of the superior breeds, which have been reared in coops, where they have been kept in such a high temperature, that they are unable to endure the open air of the country, unless in the summer season: and in the cold season, if immediately turned abroad with hens, they are liable to become *aguish*, torpid, and totally useless. The only method of safety in this case, is to keep such a cock in the house, upon the best and most nourishing food, turning the hens to him several times in the day, and permitting him to be abroad an hour or so, the weather being fine, until in a few weeks he shall be accustomed to the air. In order to keep up an efficient breeding stock, it is necessary that fowls should be reared, or fresh blood introduced, at farthest, every third year.

In making the nests, short and soft straw is to be preferred, because if the straw be long, the hen will be liable to draw it out with her claws, and with it the eggs. The hen, it is ascertained, will breed and lay eggs without the company of a cock; of course such eggs are barren. According to Buffon, a hen being properly attended by the cock for a few days, should she then be separated from him, the eggs laid by her during a month thereafter would be fruitful; but the longest period during which eggs have been hatched, under similar circumstances, as far as we are aware, has been three weeks.

It is doubtful, according to Mr. Waterton,

whether the cock *Pheasant* pairs with the hen when at liberty: we think it never *pairs* but from necessity, when it has but one hen to associate with. The progeny between the pheasant and common fowl are of course Mules, as proceeding from different species, although of the same genus. They may be obtained, with some little difficulty, which they scarcely repay, as being an improvement neither in form nor goodness of the flesh.

The *Pea-fowl* is generally kept by the London dealers, and a breeding stock, or a sitting of eggs for hatching, can almost always be obtained by parties in the country. The birds do not arrive at maturity until the third year: although, when about two years, the male begins to assume the characteristic tail-plumage. The cock requires from two to four hens to be mated with him; and when the locality of their habitation agrees with them they are passably prolific.

It was formerly universally believed that to breed *Turkeys*, it was of course necessary to keep a cock to associate with the hens; and from their gluttonous, and therefore expensive, habits, this deterred many from keeping them: but the poultry-wives of Norfolk long since discovered, that a daily intercourse is not necessary; and that if the females are placed with the males for a short time previously to the exclusion of the eggs, the whole batch of the sitting they afterwards deposit will be fecundated : it is therefore a common practice in that county, for Turkey hens alone to be kept, and when the season arrives, a neighbouring cock is borrowed for the purpose required. Where they are bred on an extensive scale, of course it would be absurd to depend for a brood upon the kindness of a neighbour. One Turkey cock is sufficient for six hens,—or indeed even more, if the practice of the above district be allowed ; but it is exposed to uncertainty, and is scarcely worth following.

It is certainly not generally known that the *Guinea-fowl* is strictly monogamous, pairing invariably : for breeding purposes it is therefore utterly useless to keep a pack of two or three hens to one male, for although from their gregarious habits they appear happy together, yet the cock never mixes with any but his mate, although without intercourse the other hens will lay an abundance of eggs—addled though they are. It is therefore absolutely necessary that a breeding-stock should comprise an exactly equal number of males and females.

The *Swans*, cock and hen, will not breed until in their second and third years respectively : they are also monogamous, pairing with as much strictness as the Guinea-fowl. For breeding stock it is necessary (as the cock and hen bird are so very much alike) to take especial care that *opposite* sexes are selected,—and to ascertain this, attention to the following points will alone assist us ;—the male has a thicker neck, and a larger " berry " at the base of the bill, than the female, and it also swims more buoyantly in the water, from its having rather larger lungs.   The maturity of both cock and hen is indicated by the size of the " berry," and the deep orange colour of that part of the bill which is of that shade.   In the regular slang of the profession, swans are styled *cob* and *pen*, to designate the sexes, male and female.

Geese and *Ducks* are good breeders: the former may be mated with advantage in the second or third year, but their progeny are always stronger if bred from older birds.   Ducks may be bred the first year.   Neither species ought to be allowed more than three or four mates, as with this limitation experience has shown that stronger and more numerous broods are produced than when a more extensive licence is given.

The *Pigeon* attaches and confines himself to one female, the fidelity of the dove to its mate being proverbial : yet notwithstanding all the sentimental trash that has been written about the cruelty of uncoupling pigeons when once matched, we have repeatedly noticed in our own loft, (where unfortunately the traps of kind neighbours rendered the circumstance one of frequent occurrence,) that the sorrowing for the loss of a mate is always speedily effaced, when another is once obtained:

we do not however recommend experimental matching and breeding, but if such be desired, or if one of a pair be lost, we do not look upon the event as of much importance, nor will the introduction of a new one be a matter of any such moment to the dove-cote arrangements as some pretend.

Young pigeons are termed "squeakers," and begin to breed at about the age of six months; of which the well-known tone of voice in the cock, just then acquired and commencing, are indications. Of course they may be matched according to the fancy of their keeper, for the purpose of varying the colours of any particular breed, &c. The best time for breeding pigeons is from April to June or July; and if they are well supplied with food, and kept warm during cold, pigeons of almost any healthy and well established variety will breed eight or nine times in the year. To preserve a vigorous breeding stock they should not be suffered to breed at the close of the season, when moulting; they must be separated not later than September, the cocks being placed together in one division of the cote, and the females in another: they should not be re-mated until the following February or March.

FATTENING.—When required for the table and not for breeding, the point for consideration on this branch of the subject is, the best mode of fattening the fowls as quickly and cheaply as possible.

The well-known common methods are, to give fowls the run of the farm-yard, where they thrive and fatten upon the offals of the stable, with perhaps some small portion of regular daily food; and at threshing time they become fat. They are also often confined during a certain number of weeks in coops, those fowls which are soonest ready being drawn as wanted. It is a common practice with some housewives to coop them for a week or two, under the notion of improving their quality and increasing their fat,—a plan, however, which seldom succeeds, since the fowls generally pine for liberty, and, slighting their food, lose instead of gain additional flesh. Such a period, in fact, is too short for them to be *accustomed* to confinement sufficiently to tolerate it comfortably. Experience shows that the best mode of fattening fowls is to place them in the feeding house elsewhere described: in that house a sufficient number of troughs, for both water and food, should be placed around, so that there may be as little interruption as possible from each other, and perches should be furnished for those which are inclined to perch, which few of them will desire, after they have begun to fatten, but which helps to keep them easy and contented until that period. Fowls may thus be fattened to the highest pitch, and yet preserved in a healthy state. We are aware, that to suffer fattening fowls to perch is contrary to the general practice, since it is supposed to bend and deform the breast-bone; but as soon as they become heavy and indolent from feeding, they will rather incline to rest in the straw; and the liberty of perching accelerates the period when they are inclined to rest on the floor.

It is better to give good sound grain in the feeding-house, putting the confined fowls upon a level with those fed at the barn-door. The water, also, given to fattening fowls should be often renewed, fresh and clean: indeed, those which have been well kept, will turn with disgust from ordinary food and foul water.

It has always been a favourite maxim among feeders, that the privation of light promotes and accelerates obesity. It may probably be so, but as a state of obesity obtained in this way cannot be a state of health, a real question arises, whether the flesh of animals so fed can equal in flavour and nutriment that of the same fed in a more natural way? Pecuniary and market interest may perhaps be best answered by the plan of darkness and close confinement; but a feeder for his own table will declare for the natural mode.

Full-sized fowls have been intended thus far; but the feeding house is well calculated for fattening the younger chickens, which may be put up as soon as they quit the hen: for generally, when well kept and in health, they will be in fine condition and full of flesh, at that period, which flesh is afterwards expended in the exercise of foraging for food, and in the increase of stature, and it may be a work of some time afterwards to recover it; this is the case with those which stand high upon the leg, which should be fattened from the hen, it being extremely difficult, and often impossible, to fatten long-legged fowls in coops. Smaller breeds and the game are the most delicate and soonest ripe. The London chicken-butchers, as they are termed, or poulterers, are said to be of all others the most dexterous and expeditious feeders, putting up a coop of fowls, and making them thoroughly fat within the space of a fort-night: using much grease, and that perhaps not of the most delicate kind, in the food.

In carrying this plan into effect, however, it has always been found practically necessary to allow a considerable number of weeks for the pur-pose of making fowls fat in coops. In the com-mon way, this business is often badly managed, fowls being huddled together in a small coop, tear-ing each other to pieces, (instead of enjoying repose,) and so irregularly fed and cleaned, that they are *stenched* and poisoned, and their flesh actually smells and tastes of it when smoking upon the table. All practical plans have their peculiar advantages; among others that of leaving poultry to forage and shift for themselves; but the best method, whether for domestic use or sale, is constant high-feeding from the beginning, whereby they will not only be always ready for the table with very little extra attention, but their flesh will be superior, in juiciness and rich flavour, to those which are put up for fattening from a low and attenuated condition. Pullets hatched in March, if high-fed from the shell, will lay plentifully through the following autumn, and not being intended for breeding-stock, their eggs may be taken advantage of in the meantime, and themselves disposed of, thoroughly fattened for the table, in the following February.

Brewers' grains, either with or without treacle and oatmeal, make an excellent fattening food, though it should not be resorted to till a short time before the fowl is ready for the table. In India fattening is accomplished very successfully and expeditiously, by a course of boiled (not too soft) rice, sago, milk, and coarse sugar. Whatever plan may be adopted with fowls under confinement, it will materially favour the process, if they are kept for the first 24 hours without any kind of food, as they will then greedily devour almost everything placed before them,—otherwise, if food is given them before they are hungry, they sometimes reject it, and seem not to care to eat after.

For fattening *turkeys* confined for feeding, sodden oat, barley, or wheat meal, mixed, is the proper food. Generally their food and treatment may be the same as for other fowls; or they may be caponized, a practice not very common here, though often adopted on the continent,—we do not, however, recommend it, as it is attended with too much risk. Almost every district has its peculiar mode of fattening—sometimes acorns, beech-mast, chestnuts, boiled and well mixed into paste with any kind of meal, are used. In Norfolk they were formerly almost wholly fatted with buck-wheat; but other plans are now in vogue: when they are six months old, they are put up to feed, when they only require a little more confinement, and abundance of food—as much as they will eat; boiled mashed potatoes, mixed with meal of buck, maize, barley, or beans, will do well, made into thick paste; at the end of a month of this diet, it is usual to give about six balls of barley-meal and sugar every night before roosting, for a week or ten days. In France the usual diet is meal-paste, mixed with chopped suet and milk, or with ale and molasses; whole pepper, garlic, aniseed, and tonic herbs are likewise given, and said to increase the savoury flavour of the flesh: the time consumed is about two months.

The following account of the mode in which *geese* are fatted by the great metropolitan poultry traders, was furnished to the author some years since by a person well conversant with the whole system :—

"Cleanliness, punctuality, and regularity prevail; the business is conducted as it were by machinery.

The fatting of geese and poultry in general, in as short a time as possible, is effected solely by paying unremitting attention to their wants; by keeping them thoroughly clean, by supplying them with proper food, (dry, soft, and green,) water, exercise ground, &c. On arriving at the feeder's, they are classed according to condition, &c.; and are fed three times a day. *Goslings*, or young geese, come to hand generally about the month of March: at first they are fed on soft meat, consisting of prime barley or oat-meal; afterwards on dry corn.

" Managed on the above mode they will be speedily fattened green, that is, at a month or six weeks old, or after the run of the corn stubbles; indeed, a goose fattened entirely in the stubbles is preferable if previously in good case, and full fed in the field. But when needful to fatten them, the feeding-houses already recommended are most convenient. With clean and renewed beds of straw, plenty of clean water, and upon oats crushed or otherwise, pea or bean meal, or pollard, mixed up with skimmed milk, geese will fatten pleasantly and speedily. Very little greens of any kind should be given to fattening geese, as being too laxative, and occasioning them to throw off their corn too quickly." It may be added that abroad, suet, oat-meal, and pot-liquor are given, mixed in a paste, with the occasional addition of a hot potatoe or two: sometimes also chandler's greaves are used, but these we object to, as giving the flesh of the poultry an oily rank flavour.

Ducks are fattened, either at large or in confinement, with plenty of food and water; or restricted to a pond, with as much solid food as they can eat,—which last we think preferable, as they fatten speedier by that mode, mixing their hard food with such variety abroad as is natural to them, more particularly if in good condition at first; there is no check or impediment to thrift from pining, but every mouthful tells and weighs its due weight. A dish of mixed food, if preferred to whole corn, may remain on the bank, or rather in a shed, for the ducks. Oats, whole or bruised, are the standard fattening material for ducks and geese, to which may be added pea-meal. The house-wash is profitable to mix up their food, under confinement; but it is obvious, whilst they have the benefit of what the pond affords, they can be in no want of loose food. Acorns in season are much affected by ducks which have a range; but ducks so fed are certainly inferior in delicacy, though the flesh is of high flavour, and far from disagreeable. Offal-fed ducks' flesh is inferior, though it does not yet emit the abominable stench which issues from offal-fed pork, and with which the dining tables of London are so frequently and satisfactorily perfumed. A favourite mode of fattening ducks is by dieting them on what butchers call " midguts," that is, the small entrails of sheep;

but all offal-fed poultry have a flavour insurmountable to our fancy,—indeed, all who have tasted the difference between what are called *town* (or offal) fed ducks, and those fatted in the country, will not hesitate in giving the preference to the latter.

Cramming.—Barley and wheat meal are generally the basis, or chief ingredient, in all fattening mixtures for poultry, whether fed in the ordinary mode or by *cramming ;* but in Sussex ground oats are used, and are in high repute for the purpose. " They are fattened there to a size and perfection unknown elsewhere. The food given them is ground oats made into gruel, mixed with hog's grease, sugar, pot-liquor, and milk ; or ground oats, treacle, and suet, sheep's plucks, &c. The fowls are kept very warm, and crammed morning and night. The pot-liquor is mixed with a few handfuls of oat-meal and boiled, with which the meal is kneaded into crams or rolls of a proper size. The fowls are put into the coop two or three days before they are crammed, which is continued for a fortnight, and they are then sold to the higglers."

The Wokingham method is to confine the fowls in a dark place, and cram them with a paste made of barley-meal, mutton-suet, treacle, or coarse sugar, and milk, and they are ready in a fortnight. It appears to us utterly contrary to reason, that fowls fed upon such greasy and impure mixtures can possibly produce flesh or fat so firm, delicate, or well flavoured, as those fattened upon more simple and substantial food, as for example, meal and milk; and we think lightly of the addition of either treacle or sugar: with respect to grease of any kind, it must render the flesh loose and of indelicate flavour. Nor is any advantage gained (excluding the commercial one) for home use, by very quick feeding: for real excellence cannot be obtained but by waiting nature's time, and using the best food. Besides all this, in attempts to fatten fowls by cramming, they seem to loathe the crams, to pine and to lose flesh; and where crammed fowls do succeed, they must necessarily be in a state of disease.

On the continent, fowls, and especially Turkeys, are crammed by means of a small tin funnel inserted in their gullet. Turkeys are crammed generally with whole walnuts, which are given every day, in numbers varying from 4 to 40, and forced down by gently stroking the throat with the hand.

It sometimes happens, whilst this *forcing* system is going on, that fowls will be unable to digest the crams fast enough between the feeding times,—it is therefore better, before cramming at night, to feel if the morning's food has been digested; if not, and the crop feels hard and full, no more food should be given then, but a little warm and thin gruel poured down the throat, which the bird will readily swallow, if afterwards the beak is held tightly between the finger and thumb. The crams themselves should be rolled as dry as possible into sausage-like pieces, not thicker than the little finger, and dipped into milk or gruel, which materially assists its passage down the bird's gullet.

Caponizing.—We have already acknowledged our inferiority in the affair of quickly feeding poultry in close coops; and have now a similar acknowledgment to make respecting Capons, never having had any success in cutting fowls, nor, in truth, much affecting the practice; which, however, has long been successfully carried on by the breeders of Sussex, Surrey, and Berks, and seems to have been almost entirely confined to that part of the country. The mode of performing the operation seems to be utterly unknown elsewhere; or granting that the common cutters and cow-leeches have some speculative knowledge thereon, they generally kill the patient, in their attempt at the practice. The reader may smile at that which may be deemed false delicacy in us, but we have naturally a kind of dread and abhorrence of all practices of this kind, however profitable. They who wish to have their fowls safely cut, where the practice is not common, must procure an operator from the proper district.

In France the poultry wives very generally adopt the plan of caponizing their fowls, which then very readily fatten at a small cost, but here a prejudice (well-founded we think) exists against the practice, on account of the risk and cruelty : and whilst disclaiming any morbid sentimentalism or affected humanity, we may be excused for not giving our readers directions for the continuance of an almost obsolete custom, not justified either by interest, utility, or necessity.

INCUBATION.—" When tyme of yere cometh, thou must take hede how thy hennes, duckes, and gese doe lay, and to gather up their egges ; and when they waxe broodye to sette them where no beastes nor other vermyn can hurte them : and thou must know that all whole-footed fowles sytt a moneth, and all cloven-footed fowles but three wekes, (except a pay-henne, and great fowles, as bustards, &c.,) and when they have broughte forth their byrds, to see that they bee well kept from crowes, and other vermyn." So wrote an old author of the 16th century, and few modern writers could give better advice at the present day,—for therein consists the whole art and mystery of successful egg-hatching ; forming, as it certainly does, a pithy summary of the art. The desire on the part of hens to " waxe broodye," or incubate the eggs, is generally indicated by an uneasy fluttering about of the bird, running and sitting in corners, by a peculiar " clucking " note she utters, and often by the comb and wattles becoming pale and rather unhealthy looking. If a hen exhibit these signs, it is better at once to place her in a sitting nest. Some persons prefer having a house exclusively devoted to sitting-hens during incubation, in order to prevent their being disturbed by laying-hens: a good plan, though not adopted in the one we suggested, which provided for the construction of sitting-nests on one side of the roosting or night-house. An admirable convenience for hatching, is an old conservatory or green-house, if such be at hand,—as the light and heat it affords are not only grateful to the hen herself, but the young ones are both hatched and reared well in it. Nothing is better for a nest during incubation than a broad shallow earthen baking-dish lined with straw cut into short lengths. Some put a small piece of flannel at the bottom, but this is objectionable from the heat and moist absorbent nature of it. The straw can be placed on the cool ground. If there is any difficulty in getting the hen to sit in the nest allotted to her, she may be confined to it by means of a board with air-holes placed over the nest,—or by means of a piece of netting stretched over the nest.

There is considerable difference in the disposition to incubate exhibited by some hens,—for whilst in many the desire to sit seems predominant, in others the desire is as slight. The desire or *furor* for sitting is sometimes endeavoured to be excited in hens by means of food: thus sops of bread and wine, leaves and seeds of nettle, mustard, dried and

powdered, are given ; but we think all such unnatural excitements or efforts to *force nature*, are
hurtful to the fowl. Leaving some nest-eggs of
white wood, or chalk, in the sitting-boxes, is an
enticement which harmlessly hastens the manifestation. On the other hand, much useless cruelty is
too often exercised, to prevent the hen sitting when
eggs rather than chickens are in request,—such as
immersing the fowl in water, thrusting a feather
through the nostril, &c. ; it is time that all such
useless and barbarous practices should meet with
the censure and contempt they deserve.

Most persons have witnessed the superstitious
formality practised by old women in setting eggs
for hatching,—always attended by some mysteri-
ous reference to odd numbers, lucky days, &c.
Eggs intended for sitting on should be removed
from the laying-nest as they are deposited, until
a sufficient number has been obtained to form what
is called a " clutch." In the meanwhile they must
be carefully put away in a deep box, set on end and
covered over in bran, the small end downwards,
and placed in a cool and dark place. For sitting,
the newest are to be preferred, and none of a
greater age than two or three weeks can safely be
relied on to hatch,—although some isolated instances have occurred of much older eggs demonstrating the pertinacity of vitality. The following
experiments, carefully recorded by a friend, are
interesting as relating to a subject on which we

believe no similar information has previously been given.

Date 1852.	No. of Eggs set.	Breed.	Age of oldest.	Result.		
				hatched	addled	died.
March 4	10	Cochin	19 days		10	
30	13	—	25	2	11	
31	11	—	15	1	8	2
April 7	11	—	18	8	2	1
12	11	—	26	3	6	2
15	6	—	10		4	2
May 4	9	—	18	5	2	2
7	18	—	15	10	5	3
March 14	11	Spanish	23		10	1
April 6	12	—	18	10	1	1
11	11	—	9	3	7	1
26	9	—	15	5	3	1
May 7	9	—	11	4	4	1
March 2	11	Poland *	41	1	8	2
April 27	10	—	11	5	4	1

Of course the differences of the seasons, temperature, &c., must have a considerable effect upon the chances of hatching a sitting of eggs. In choosing eggs for the purpose, they should be selected as nearly as possible of the same size, and free from any circular flaw, roughness, or cracks, in the shells.

The number of eggs proper to place in the nest varies from nine to fifteen, according to the season of the year and the size and capabilities of the sit-

* It is worthy of remark, that the newest egg of this clutch had been laid on the 30th of the preceding January; therefore, supposing that this identical egg was *the* one that was hatched, even then it must have been 32 days old.

ting hen; and it is better to put in too few rather than too many; as if more are placed under the hen than her natural heat can well incubate, half will be addled, and the hen herself injured. A good plan in cold weather is to put eggs, before placing them in the nest, wrapped in flannel, near to the fire, until they have attained a warmth of about 75° Fahr. (the usual warmth of a hen's body), as the sitting hen, when she covers them, will then not have to lose so much of her own heat. Odd numbers are generally considered preferable upon the supposition that they lie closer in the nest— the odd one being in the centre. The eggs are to be marked and inspected when the hen leaves her nest, to detect any fresh ones laid, which should, of course, be immediately taken from her. The box and nest must be perfectly clean for the hen, and a new nest not thrown upon an old one, from the filth of which vermin are propagated. Eggs broken in the nest should be cleared away the instant of discovery, and those remaining washed with warm water, and quickly replaced, lest they adhere to the hen, and be drawn out of the nest; if necessary, the hen's feathers may also be washed with warm water. A foolish practice is often recommended of *turning the eggs when the hen gets off the nest*, in order that every side may be properly warmed; this, however, must be studiously avoided,—the hen has a natural instinct which renders her a far more careful and intelligent incubator than aspiring humanity can design. An equally reprehensible plan is that of dipping the egg into warm water upon the eighteenth day of incubation, upon the absurd fancy that it would assist hatching. With respect to the capriciousness of some hens in sitting, it is a risk which must be left to the judgment of the attendant, as to whether or not the hen which appears desirous of sitting may be safely trusted with eggs. Very often the hen will cluck and appear anxious to incubate, and yet, after sitting on the eggs given her, just long enough to addle, will desert them.

Corn and water should be placed beside the sitting hen, whenever necessary, to encourage steady incubation, and to support those in which the natural excitement is so powerful, that they will remain several successive days upon the nest. The plan of feeding on the nest should be invariably pursued with all frequent sitters.

Pheasant.—However good nursing-mothers in a wild state, pheasant hens are far otherwise in the house, whence their eggs are almost always hatched at home by the common hen—usually a bantam; but certainly the pheasant is otherwise far the more eligible, on all accounts. The natural nest of the pheasant is composed of dry grass and leaves, which, being provided for her in confinement, she will sometimes properly dispose. The cock is voracious and cruel, for which reason it is necessary to watch them carefully in confinement, to ascertain if they exhibit any such indications ; for if so, it will be vain to hope to rear a young brood.

Turkey.—The hen will cover from nine to fifteen eggs ; and, unless attended to, will, perhaps, steal a nest in some improper or insecure place abroad.   On this account, where Turkeys are bred to any extent, and are permitted to range, it is necessary to allow them a keeper.   The hen is, nevertheless, extremely vigilant and quick in the discovery of any danger that may threaten her brood.   The Turkey hen often manifests her desire to sit so strongly that she will even sit upon stones if not allowed eggs ; and it is therefore much better and more humane at once to place them in the hatching-nests, in a dry, warm, and rather secluded corner, formed with a little straw, and a pad or mat of the same placed over it.   The most desirable time for *setting* Turkey hens is from the middle of February to the end of April : and if any difficulty exists in obtaining a sufficient number of eggs in time, a few common fowls' eggs may be put with those of the Turkey,—the former being set *one week after the latter* have been covered, to allow for the difference of time required for the respective incubations.   Although an admirable sitter, the Turkey hen is excessively shy and timid, and must upon no consideration be intruded on whilst at the nest, except by the regular keeper by whom she is customarily attended and fed.   It is also desirable not to allow the Turkey cock access to her presence whilst she is sitting, for at that time he will sometimes tear the hen from her nest, and, in the struggle, destroy the eggs.

Guinea Fowl.—The wild and rather untractable habits of these fowls, render it a matter of difficulty to attach them sufficiently strongly to any particular nest or spot in the poultry-house, they preferring generally to stray over the grounds, until behind some bank-side or hedge-row they select a nest for themselves. It is more profitable to hatch Guinea fowls' eggs under common hens, as they cover better and also a greater number. Game fowls and Bantams are sometimes employed in the process; but there is some risk even in that, as the time of incubation is longer for the Guinea eggs than for the common, and sometimes the common fowl will tire before the period is quite up, or at least will not sit so steadily the few last days as is required.

The *Swan* is a bird of courage equal to the pride it apparently possesses; and both the male and female are extremely dangerous to approach during incubation, or whilst their brood is young. They both labour hard in forming a nest of water-plants, long grass, and sticks, generally in some retired part or inlet of the bank of the stream on which they are kept. The hen begins to lay in February, producing an egg every other day, until she has deposited seven or eight, on which she sits six weeks, although Buffon says it is nearly two months before the young are extruded. When the materials for constructing a nest are not *naturally* at hand, it is expedient to provide a load or two of sedges and rushes to be laid on the banks, which is all the trouble entailed on the proprietor in nest-making,—the birds performing the rest.

Goose.—During the breeding season, in the Fen counties, the geese become the joint tenants with their proprietors in their cottages. When, however, they are reared in the poultry-house, a nest should be prepared for the goose in a secure place, as soon as she exhibits a desire to incubate, which she does almost immediately upon commencing laying. The earliness and warmth of spring promote the early laying of geese, which is of consequence, since there may be time for two broods within the season. However, to attain this advantage, it is necessary to feed breeding geese high throughout the winter, with solid corn ; and on the commencement of the breeding season to allow them boiled barley, malt, fresh grains, and fine pollard, mixed up with ale, or other stimulants. The goose covers readily from eleven to fifteen eggs. With respect to feeding the goose or duck upon the nest, it may be occasionally required, but is not a thing of much account, since they will generally repair to the water sufficiently often, from their natural inclination.

The *Duck* will cover the same number of eggs
as the goose.   When she first begins in February,
she will often, unless watched, lay her eggs
abroad, and, concealing them, will incubate out of
doors.   She generally lays by night, or early in
the morning, seldom after ten o'clock, except in
cold weather, when she will occasionally retain her
egg until mid-day or afternoon.   To know when
to keep her within until she has laid, her appearance and weight behind may be trusted to by constant observers : once accustomed to a nest, she
will not forsake it.   The duck swimming with her
tail flat and level with the water, indicates her egg
ready for protrusion.   It was formerly directed to
give each duck her own eggs to incubate ; experience shows that not much consequence need be
attached to the point, though no harm can come
of attention to it.   During incubation, the duck
requires a secret and safe place, rather than any
attendance, and will, at nature's call, leave her
eggs to seek her food. Ducks' eggs are often hatched by hens, when ducks are more in request than
chickens ; and the plan has no objection in a confined place, and with a small stock, without the advantage of a pond ; but the hen is much distressed
at witnessing the supposed perils of her children
venturing upon the water.   We have heard of setting duck's eggs under a goose, which would, of
course, cover a considerable number.

Pigeon.—The pigeon, it is needless to state, is

monogamous ; and after being matched about a week or ten days the hen will produce her eggs—two in number : all writers assert that she rests one day between her first and second egg, but this is by no means invariably the case with good breeders and under favourable circumstances. Our own loft (comprising principally Antwerp Carriers) afforded *numerous* instances of eggs laid on *two consecutive days*, and with periods intervening varying from twenty-seven to thirty-five hours. It has been recommended to remove the first egg, to place in the nest when the pair is deposited,—this we object to, as some pigeons will not incubate after they have been disturbed ; if permitted to stand guard over it, no harm will generally happen from the pigeon treading on it. If matched for the first time, it has been our practice not to let the pigeon sit over the first eggs, as they rarely are fertile : they will sometimes produce three eggs before incubating, in which case one should be destroyed. Before laying, a broad earthen circular pan (easily obtainable at the earthenware shops) should be placed in the cote, with a few wisps of straw cut short, and the harshness removed by rubbing between the hands, strewed about, which the pair amuse themselves by forming into a suitable nest : occasionally they prefer making the nest upon the ground, which may be permitted if the weather is warm : if a pan is used, a brick or foot-lodge must be put at the side to prevent accidental breaking of the eggs by the birds descending from the perch or alighting suddenly upon the nest. The cock and hen share in the process of incubation,—the former invariably sitting over the eggs from about nine o'clock in the morning till four or five in the day, when the hen presides for the remainder of the twenty-four hours.

INCUBATION.—The duration of incubation over eggs, for the purpose of developing the embryo contained within, varies considerably in different species. The time which is consumed by the different varieties of fowls is materially affected by the state of the temperature, and the steadiness or otherwise of the incubator; and the exact time at which the eggs are set should always be recorded, and subsequently the time of hatching. By this means many interesting facts at present unknown, because unobserved, will be ascertained.

The term of incubation for the common domestic fowl is from twenty to twenty-one days, a few hours earlier or later, according to the influencing circumstances we have already alluded to; but there is occasionally a deception on this point, —for some hens *stand over* the eggs during the first two or three days,—thereby inducing an erroneous idea in the end, that they have sat some days beyond the usual period.

The pheasant occupies precisely the same time as the common fowl.

The peahen sits from twenty-seven to thirty days.

The Turkey, Guinea-fowl, swan, goose, and duck usually occupy about four weeks; but they are very uncertain, often varying from twenty-seven to twenty-nine days, occasionally prolonged to thirty-one, and even thirty-two, days. Among ducks, especially, the period required for incubation differs considerably in the varieties: thus it is said upon competent authority, that the Muscovite rarely sits less than five weeks, and even that unusually long time has been occasionally further extended to thirty-eight and forty-two days. With the China goose also incubation lasts five weeks, generally.

The pigeon consumes, jointly with the cock, seventeen or at most eighteen days in the process; although this time is usually stated at from seventeen to twenty days: careful observation in our own loft, gives the seventeenth complete day *after the second egg is laid*, as the almost invariable time for hatching.

HATCHING.—Upon the expiration of the term we have indicated, the process of the actual hatching the chick must be carefully watched, the warm or cold state of the weather making, as we have said, some hours difference. In a work which appeared very many years ago in France, Reaumur described fully, and, with some important exceptions, tolerably correctly, the position of the embryo chick in the egg, and its progress, from the commencement of its development towards incipient existence, up to the moment of its hatching, and to that source (whence they have been copied and re-copied times without number) we must refer the reader for the details. A singular error was, however, promulgated by that eminent and scientific man, which, strange to say, up to a recent period, never sufficiently engaged the attention of naturalists, to lead to its discovery and controversion, and it has therefore been implicitly believed to be a fact, from Reaumur's days to our own. The chirping, ticking, or tapping kind of noise, often heard for some hours (from 30 to 40) before the chick emerges from the prison-house in which it is confined, led Reaumur to suppose that it was caused by regular taps or blows inflicted by the bill upon the shell, by which the latter was fractured to a sufficient extent, to admit of the chick's passage out; and certainly its feasibility may account for its general acceptation. To F. R. Horner, Esq., M. D., of Hull, the scientific world is indebted for clearing up this long-established error; he having, after devoting considerable time and attention to the whole subject, ascertained that the noise heard in the hatching egg is not caused by any tapping or contact of the bill of the chick upon the shell, as commonly supposed,—but " is truly respiratory, and produced by the transmission of air through the lungs—or, in other words, that it is the natural respiratory sound of the chick." The correctness of this account, Dr. Horner indisputably demonstrated in a series of experiments detailed in a most interesting paper, read by him before the British Association for the Advancement of Science. We much regret our inability, for want of space, to insert this memoir, though kindly placed in our hands for the purpose.

Nature, as Reaumur long since observed, has committed to the chicken itself the task of breaking its way through the shell, when the proper time has arrived; but all chickens do not perform that task in equal time, some appearing, from weakness, or want of heat, to be unable to disencumber themselves of the shell. In such cases, at or near the expiration of the full time of nature, if the shell is not at all, or only partially, fractured, assistance may sometimes be rendered successfully, but in this, caution, skill, and scientific knowledge, which few possess, are indispensably necessary. Generally, the chick with one blow makes a circular fracture at the larger end of the egg, and a section of about one-third of the length of the shell being separated, delivers the prisoner, pro-

vided there be no obstruction from adhesion of
the body to the membrane which lines the shell,
which, however, sometimes happens; in which
case the method of assistance is, to take the egg,
and, dipping the finger, or a pencil-brush, in warm
water, apply it to the fastened parts, until they are
loosened, by the gluey substance being dissolved
and separated from the feathers; the chick then,
being returned to the nest, will extricate itself, a
mode generally to be observed, since violence used
would often be fatal.   Sometimes, however, it may
be necessary to break the shell, by gently striking
it with a key, and then removing the membrane
as softly as may be from the feathers, by means of
the fingers only,—the moisture being also applied
as above.   In such cases, however, we have had
but little success attending our efforts to aid nature; and, upon the whole, the advantages of
giving assistance under any circumstances is doubtful, as they rarely make *strong* birds, and very
rarely even *healthy* ones.   A far preferable plan
to adopt, when the proper time is passed, and two
or three eggs, apparently with life within, remain
unhatched, is to remove them, and gently " springing " or cracking (but not breaking) the shells,
place them under another sitting hen whose time
has not expired, when the increased warmth of
body will often produce vitality; and, at any rate,
the eggs being removed from the first nest, will

prevent the young chicks which have been there hatched, from sustaining injury by contact with the shells of over-due hatching eggs.

Artificial hatching being now carried on more extensively than formerly, we may be expected, perhaps, to say a few words upon that subject, as it is impossible to say to what extent or completeness it may ultimately be carried, though, for our own part, we see most insuperable objections to its ever being successfully adopted.

In the previous editions of this work, the Egyptian and other modes of artificially hatching fowls' eggs were fully described and dilated upon; and also some of the earliest experiments made in this country, conducted by the author himself whilst residing in Surrey, in 1782. He was, however, soon satisfied with the trial upon a very small scale, and abandoned the scheme; for although perfectly satisfied with his success in hatching a considerable number of eggs, there was no adequate motive to pursue it in this country, where a quantity of poultry, fully equal, and even superior to the demand, may be raised by the natural means: were it otherwise, no doubt the artificial process might be conducted with sufficient success in the multiplication of domestic fowls. No person will then attempt artificial hatching, but from the motive of mere curiosity, and that motive must indeed be powerful, to carry one through the endless labour and attendance required.

Hatching artificially by means of steam succeeded; and in 1823, various attempts were made at London, Bath, and other places, to bring the plan into general use, but the movement was unattended with any practical results. More recently a Mr. Cantelo has perfected and successfully brought out an ingenious machine for artificial hatching, called a " Patent Hydro-Incubator," which almost infallibly hatches all good eggs intrusted to it—but (and that is everything) the *machine will neither feed nor rear* the young when produced, and their exit from the world is rather more rapid than their entrance. The shilling exhibitions of this last-mentioned machine, has rendered the public fully acquainted with the process by which the work is accomplished; but, for various reasons, the plan has never yet been rendered of any public or general utility.

REARING.—The chicks once ushered into the world, all that is necessary, is to take the first hatched from the hen, in order that she may quietly complete the incubation of the eggs still hatching. As they are removed they may be secured in a basket of wool or soft hay, and kept in a moderate heat; if the weather be cold, near the fire. They will require no food for many hours, even four-and-twenty, should it be necessary to keep them so long from the hen. A tiny morsel or crumb of bread wetted with ale, and very gently placed in the mouth of a newly-hatched chick, will not hurt it, but gin, rue, pepper-corns, and all such nonsensical medicinal dieting, must be carefully eschewed, being little better than so many modes of killing them off. Inspiration of air and absorption of heat, appear to be the only sustenance chicks at first require. At night, or before, if the remaining eggs are hatched, the young ones should be replaced under the hen. The whole brood fairly hatched, they may be placed for the first two or three weeks in a large coop, boarded on any exposed sides and lathed in front; the best place for such a coop is abroad upon a dry, sheltered, and sunny spot, if the weather be warm,—if otherwise, the whole must be confined within doors : in either case the coop, if left open, (which is generally advisable after the first few days,) should not be set within the reach of another hen, since the chickens will mix, and the hens are apt to maim or destroy those which do not belong to them : nor should it be placed near numbers of young fowls, which are likely to crush the straying young chicks under their feet, being always eager for the chickens' meat. The coop may be of any form dictated by convenience and materials at hand : a good and simple one is constructed of two wide boards nailed together at the top only, in the shape of a model pyramid, with a wooden back or end, and the other end lathed with moveable bars of wood, capable of being pulled up to admit the chickens going in or out if allowed so to do. A grocer's common tea-chest may also be employed, one side being knocked out, (so that the coop has no bottom next the ground,) and the front lathed as in the trian-gular-shaped coop. Generally speaking, and de-pendent on the weather, situation, and disposition of the hen, there is no necessity for cooping the brood beyond two or three days, as they are much benefited by the scratching and foraging of the hen. They must not be let out too early in the morning, or whilst the dew remains upon the ground, far less be suffered to range over the wet grass, one common and fatal cause of disease. Another caution is to guard them watchfully against sudden unfavourable changes in the wea-ther; for *nearly all the diseases of gallinaceous fowls arise from cold moisture.*

For the period of the chickens quitting the hen, and shifting for themselves, there is no general rule: the most certain is, when the hen begins to roost, leaving them; if sufficiently forward, they will follow her; if otherwise, they should be put in a proper place, with other young poultry, as nearly of their own age and size as possible.

For the first food there is nothing better than eggs boiled hard, stale crumbs of bread, curds, and a small quantity of split groats, all macerated and cut up very small, and given *as often as the chicks will eat,* which is the best rule for feeding, stated times falling short of their requirements; for young chicks, like young children, according to the old rhyme,—

" Would always be stuffing."

As the chicks have not only to supply the growth of their bodies, but also to contribute a large portion of the sustenance they take towards the production of feathers, the necessity of supplying them with a superabundance of nourishment is apparent: and in spite of what has been said of the danger of "over-feeding" chicks, the rule for feeding may safely be left to natural instinct alone, as they are more likely to eat beyond their requirements, when fed only once in a while at stated hours, than when the food is always before them, to eat when nature dictates. After a short time split groats may be given in greater proportion; small wheat and barley may also be thrown down, and if too large the mother will generally break them for her brood: later still, pearl barley simmered tolerably soft, and rolled up in dry meal till rather stiff, forms an admirable change. All vegetables, sloppy or watery food, such as potatoes, soaked bread, &c., however admirable in after dieting, should be scrupulously avoided as highly improper. When older, tender cabbage and lettuce leaves, and, indeed, any kind of good vegetables, may be permitted; and oatmeal, rice gently boiled but not broken up into a mash, with stale bread steeped in warm new milk, (quite fresh and sweet,) as almost all sorts of good food, (except animal,) may be freely given. For all fowls, but especially when young, it is perfectly indispensable that their drinking water should be pure, clean, and often renewed; and there are convenient pans or troughs made in such forms, that the chickens may drink without getting into the water, which often injures them: a basin whelmed in the middle of a pan of water will answer the end, the water running round it.

In reference to the rearing of *winter* chickens in this climate, it will always be found difficult, and rarely successful, even in a carpeted room, and with a constant fire,—generally, we think, the attempt will scarcely be worth the trial. To essay to raise winter stock, except within doors, can scarcely be expected to succeed in the least, even upon dry soils and in the best situations; and upon clayey or damp sites it is utterly useless.

It is usual to give numerous special directions for rearing the young of the different varieties of our domestic fowls,—this we regard as gratuitous and unnecessary: it is a vulgar saying that " eggs are eggs all the world over," and assuredly it is equally true that *chickens are chickens*, of whatever breed they may be; and, with a few solitary exceptions, the management and food requisite for their successful rearing should be the same. The following suggestions seem alone specially called for.

Game chicks are hardy, but as their nature requires rather higher breeding than other fowls, a little hemp seed, and black pepper gently dusted over their first food, may be occasionally given them perhaps with advantage. They seem better capable of early shifting for themselves; as, with high feeding and a good attendance of cocks, the hens, if permitted at large, will frequently begin laying when their broods are only three weeks old.

Dorkings are regarded as by no means hardy when chickens, and so require most careful management: assuredly they are more pre-disposed to diarrhœa, or scouring, and for that reason the food given them should not be in any degree sloppy or watery, and a greater proportion of eggs and rice must be mixed for them.

Spanish chicks seldom are *perfectly* feathered till approaching full growth, though the pullets attain their plumage sooner than cockerels: it is most advisable to hatch Spanish eggs, therefore, as early as is prudent in the spring, perhaps about the end of April, in order that they may be properly feathered and thereby protected when the autumnal cold comes. The cocks generally obtain good white faces much sooner than the hens; but this is often exceedingly tardy in making its appearance, and for this reason it is extremely difficult to select a good specimen when young.

Poland chicks need (if possible) greater warmth and dryness of situation than the young of most other kinds. There is great difficulty to distinguish the sexes until they have cast their first feathers. The spangled varieties generally improve in appearance with each moulting, and never get their full beauty of plumage till the second year.

Bantams cannot be reared better than by allowing them, for the first few weeks, the unlimited range of an old hot-bed, for they are inordinately fond of minute and destructive insects, which are the terror of our agricultural friends; some of whom, writing in the columns of the *Gardener's Chronicle*, urge upon farmers the keeping of Bantams, for the serviceable aid they render in getting rid of wood-lice and other vermin from the cucumber, melon, and strawberry-beds. The usual time for hatching is May; but, in order to dwarf the proportions, they are often raised in the autumn.

Pheasant.—The following original and practical instructions for rearing pheasants, were communicated to the author, many years ago, by Mr. Carstang, the eminent breeder:—If the eggs do not hatch well together, treat the young pheasants as directed for common chicks; and when the brood is all hatched, coop them with the hen, with whose call they will soon be acquainted, and may then have their liberty to run on the grass-plot, or elsewhere, observing to shift them from time to time *into* the sun, and *out* of the wind; also not giving them liberty till the sun is up, and always shutting them in with the hen before it sets. The birds, if hatched early, may be so treated until August or September, or if late, even till December. Before they begin to shift the long feathers in the tail, they are to be shut up with the hen at night; but when they shift, they are not willing; those intended to be turned out wild, should then be encouraged to perch. The country then being covered with corn, &c., they will speedily shift for themselves. The most convenient season, therefore, for setting pheasants' eggs, will evidently be April.

Young pheasants intended for breeding stock, or to be turned out in the following spring, must be provided with a new place, large enough for two pens, where none have been kept before; and there put them as they begin to shift their tails: those intended to be turned out at a future time in one pen, netted over, and those for breeding in the other; cutting one wing of each such. The Gold and Silver Pheasants pen earlier, or they will be off: cut the wing often. First food, boiled egg minced, boiled milk and bread, ants' eggs, and alum curds,—a little of each sort, and often. Alum curds form a most valuable article of food for these birds, and also for poultry: Mr. Carstang's plan was to boil a moderate-sized piece of alum in new milk, to a custard-like consistency, and a little given twice a day, with ants' eggs, after every time they have had enough of other food: if they do not eat heartily, ants' eggs will create an appetite, but they must not be given in such abundance as to be considered their regular food. Fine dry sand or gravel should be strewn near to the young birds, as they greedily eat it with their food, which may sometimes be thrown down in the sand to amuse them. Ants' eggs being scarce, hog-lice, ear-wigs, or any insect may be given; or employ artificial ants' eggs of flour beaten up with an egg and shell together, and rubbed into small pellets between the fingers; or a few "gentles" may be given: these are obtained from good liver tied up, the gentles, when ready, dropping into a pan or box of bran beneath,—to be given only very sparingly. Tender green vegetables may, after a short time, be cut up and mixed with the usual food. When first *penned* or confined, all young pheasants should be fed with barley-meal, dough, corn, and plenty of green turnips. It was formerly held that they could not be reared on any other food than ants' eggs, of which a supply could never be depended on; but the above will be found sufficient substitutes.

Peafowl.—Until about six weeks, peachicks may be cooped with the hatching-hen, whether peahen or Turkey, and treated similarly to Turkey chicks—except, perhaps, that the preference they exhibit for barley and young vegetables, calls for a larger quantity of that description of diet. But for the young being lost, or straying, they would thrive much better if suffered to range at pleasure when a week old, as they do not like confinement.

Turkey.—It is an old custom to withdraw Turkey chicks from the nest as soon as hatched, plunge them into cold water, and then wrap them in wool or flannel, giving them a whole pepper-corn and teaspoon-full of milk: this practice we cannot too strongly condemn,—it is analogous to inoculation for the small-pox—generally worse than the disease. Left with the hen for the first twelve hours of their existence, they need no food: they must be carefully housed for the next month or six weeks, dependent upon the weather; if fine, warm, and dry, no better shelter need be desired than a coop on a grass-plot, and housed at night. First food, small crumbs of bread and curds, or hard eggs chopped up; also, oat or barley meal kneaded with milk, and frequently renewed with *pure clear water rather than milk,*—our preference for water over milk (so much recommended by old writers) arises from the observation, that chickens at large among the troughs of milk-fed pigs generally are sickly and scouring, and rough in their feathers. If the chicks appear sickly, and the feathers ruffled, indicating a chill from some cause, half-ground malt with the barley-meal, and (by way of medicine) powdered caraway or coriander seed, may be given. Boiled meat, pulled into small strings, forming, as it were, *artificial worms,* may be thrown about, though not too freely. A fresh turf of short sweet grass provided daily, cleared from snails or slugs, (which will scour young chicks,) is very good for them. The above substantial food renders it quite unnecessary to waste time in collecting ants' eggs or nettle-seed, or to give clover, rue, or wormwood, as old house-wives direct. Eggs boiled hard are equally pro-per with curd, and generally nearer at hand. Boiled (not mashed) rice, mixed up with cress, lettuce, and onion-green, with occasional dishes of butter-milk, meal, melted fat, cheese shreds, and minced liver, as the chicks get older and stronger. But the grand secret of rearing them success-fully, (and, indeed, almost every other kind of fowl,) is constant and liberal feeding,—never let their feeding-platters be empty; nature is a good guide for the young chicks; and few, indeed, can give that constant attention to feeding them by the hand, as would otherwise be required. At two periods of their lives Turkey chicks are very apt to die from want of due care and attention,—namely, about the third day after they are hatched,—and when they shoot what is called the *red head*, at about six or eight weeks old: at the latter period, a few old beans, split small, may be mixed with advantage in their food.

The weather being favourable, coop the hen abroad, about two hours in the forenoon, in a moderately warm sun, whilst the chicks are only three or four weeks old, great care being taken that they do not stray from the coop. Six weeks is their longest period of confinement, after which it is more safe to coop the hen for another fortnight, that the chicks may acquire strength abroad, sufficient to enable them to follow the dam. When half grown and well feathered, they become sufficiently hardy; and, in a good range, will provide themselves throughout the day, requiring only to be fed at their out-letting in the morning, and on their return at evening; the same in spacious farm-yards: if confined to the poultry-yard, their food and treatment is similar to that of the common cock and hen. Care must, however, be taken not to let the hen and brood out too early, or before the latter are tolerably strong and hardy, as the hen, being a great traveller herself, will thoughtlessly drag her chicks over field, heath, or bog, never casting even a look behind her to call in her straggling young,—nor yet scratching for their food, leaving them entirely to their own instinct and industry for their daily supply.

If any notion is entertained of a second hatch, the sooner one hen is turned away and her brood mixed with that of another hatched about the same time, the better chance there is of rearing it; as the hen which is so turned away will generally hatch again ere July is out.

Guinea-fowl.—No special directions are needed for rearing these, as they thrive well under the same treatment as Turkey chicks,—except, perhaps, that they require even more liberal feeding, and should not have the larger grains of corn or barley given them until about half grown.

Swan.—Cygnets, as soon as they are strong enough, are led from the nest to the water by the old ones, where they live and feed on the weeds and infusoria: if, however, the water does not abound with such food, a few handfuls of oats, or corn, may be thrown in, and also a tuft or two of grass, as the young have not always the sense to come to the bank to crop it. A few *small* fish, shrimps, spawn, &c., placed in the stream, give them a rich treat, as they will dive eagerly for such food. A feeding trough, placed in a secluded part of the bank near the nest, will often tempt them to the bank.

Goose.—When first hatched, it is not advisable, nor, indeed, very practicable, to take any of the gulls from the goose, as she might kill some in her struggles to defend them. It was formerly recommended to keep the goose and brood in-doors for the first week,—but that is unnecessary: they may be penned between four hurdles upon a plot of dry grass, well sheltered, putting them out late in the morning, and housing them early in the evening: for convenience several broods may be parted with more hurdles, but this is not absolutely required, as they are generally sociable and harmless. The first food, barley-meal, bruised oats, or fine pollard, mixed up with water, and *some* chopped and cooling greens, of which none are better than the common *goose-grass*, (popularly called cleavers,) which makes its appearance in hedges about the time gulls are hatched. Some breeders state they have mixed bean or pea meal in the ordinary food, after a week or so, with advantage.

After the young become pretty well feathered, they will be too large to continue brooding under the mother's wings, but will soon sleep in groups by her side, and must then be supplied with clean straw beds. Being now able to frequent the pond, and range the common, they will obtain their living; and few people allow them any thing more, excepting the vegetable produce of the garden.

Duck.—Upon hatching, the duck may be suffered to retain the young in the nest her own time, barring accidents: on her moving with the brood, a coop may be provided, if fine, upon the grass, or if otherwise, under shelter,—but in either case placed apart from any other. The confinement is not generally necessary beyond a fortnight; and if they are strong and the weather warm, they can often be let out to enjoy the pond (the shallower the better at first) at the end of a week, but they must not be allowed to remain in too long at one time. At first it is usual to clip the down from under the tails of the ducklings, since in draggling in the water they are apt to get affected by the wet. The first food is precisely similar to that of young goslings: though it is a common practice to feed them for the first few days upon small pellets of flour or meal, rolled or rubbed up with egg, or simply with water. A wide, flat, and shallow pan of pure water (often renewed) must always be at hand.

The old wife's plan of suffering a hen to hatch a chicken or two with the ducklings is most unwise: the hen, for the sake of her natural progeny, will neglect the ducklings; whilst the aquatic nature of the latter will be always urging them to the water, whence they return cold and half drowned, and their mother perhaps unwilling to brood or warm them.

Pigeon.—Young pigeons (" squabs " they are called by the Fancy) when hatched must be left in the nest; handling is always hurtful, and will even sometimes induce the old ones to desert them altogether. The parents save the proprietor all trouble in rearing the young, as from the first they are fed and nourished by the old birds: the mode of feeding is curious; the squab thrusts its bill into the side of the mouth of one of the old birds, which forthwith pumps up the contents of its stomach (called " soft meat," technically) into its own mouth, thereby converting it into a trough or reservoir of half-digested food, which the young thence sucks down through its own bill.* This

* Mr. Yarrell (*Brit. Birds*, vol. ii. page 280) has fallen into a singular error, completely *reversing* the mode in which the pigeon

" soft meat " is usually described as " a liquid pap, prepared instinctively in the stomachs of the parent birds just before their young are hatched;" and we shall perhaps be considered bold in venturing to dissent from this popular and invariably credited explanation: we believe it is nothing more than the half-digested food eaten by the old birds and pumped up when the young require feeding. Careful observation confirms that opinion; as oftentimes we have seen the old birds feed themselves at the hopper of peas, and after a draught of drink apparently to soften and assist the digestion, in a few minutes proceed to pump it up as food for their young, and that too long before the latter could by possibility take any *hard* or *solid* food: again, we have never observed the old birds " sick " (as most books assert they are) when the young have died, and which, if the " soft meat " really were a *special secretion* of the stomach, they would of necessity be when they could not get rid of it. Gradually as the squabs get stronger, the food is injected by the old birds in a more solid or less digested state, until at length the peas or grains are thrown up from the parents just after they are taken in, when the young shortly commence pecking their food from the hopper for themselves. If the parents neglect their young, (as is often the case with Almond

feeds its young, by stating that the old birds *insert their bills into those of the young squabs.*

Tumblers and some other fancy sorts,) it is absolutely necessary to " shift " them to another nest
of more steady nurses; for which purpose it is
usual to set several common kinds of pigeons
about the same time or shortly after the more
valuable, and then to destroy the young of the
common, and give them the fancy squabs to rear.

When the young seem sufficiently strong to
bear a change, the old straw should be removed
and a new and clean nest supplied. As soon as
the young can shift for themselves, it is better to
remove them altogether, as if they remain and the
parents again sit, they will frequently, by running
to the old birds on the fresh nest, cause the eggs to
be either crushed or addled by unsteady sitting.

When the old birds are lost or die, the young
may be reared by hand; the amateur masticating
the peas or grains, and placing the bill of the
squabs in his mouth, they may generally be successfully fed: but unless they are very valuable
birds indeed, they are not worth such a tax upon
time and patience,—to have common feeding
pigeons always at hand, is far preferable as a substitute for the natural parents.

FEEDING.—An idea prevails with many, that
any sort of corn will do for poultry,—this is a
grand mistake; those who feed largely know
better: and it will be found most advantageous to
allow the heaviest and best, which shows itself not

only in the size and flesh of the fowls, but also in the eggs. Barley and wheat are the great dependence for poultry: the heaviest oats will keep them, it is true, but neither go so far as other corn, nor agree so well with the chickens, being apt to scour them. Brank or French wheat is also an unsubstantial food. Oats are recommended to promote laying; and in Kent, Sussex, and Surrey are deemed superior for fattening. Sunflower seed periodically has been recommended with high commendations, but never yet attended to by the generality of feeders: in small quantities the experiment may easily be made. A porridge of rice and barley-meal, (in equal quantities,) with millet seed added, and given with the customary food alternately, forms an excellent variation in poultry diet.

Poultry which have their fill of corn will eat occasionally cabbage or marigold leaves greedily; as also the tops of turnips, lettuce leaves, boiled and mashed potatoes, tufts of good fresh grass, &c.; and they should be well supplied with green food,—without which they will not enjoy good health, the flesh not so firm, and the eggs poor, the yolks being of a pale, sickly colour,—indeed nothing seems to have so much effect upon the eggs, as the green food a fowl eats; for which reason we strongly object to leeks and onions being given to laying fowls, as the eggs generally partake of the flavour: and the same may be said of chandler's greaves,—no kind of food imparting worse savour alike to eggs and flesh. Fowls, however, though by nature graminivorous, are also great devourers of flesh-meat, and kitchen refuse may therefore *occasionally* be given, but only when in a *wholesome* state, and in *very small quantities :* for insects and animal food (sparingly) form part of their natural diet, and the entire want of such will perhaps impede their thriving. Malted or sprouted barley has been thought to have a good effect in promoting laying, as also cordial horse-ball in the cold season, at least so old wives say, but we should not advise a long continuance of such diet.

As to specific quantities of food to be allowed, it is difficult to fix any; the phrase, a "small quantity," may mean *ounces* or *pounds*, according to the kind used or the fancy of the feeder: it is a case in which precision ought not to be expected, from the various circumstances of size, age, season, nature of the food given, &c. For successful feeding and rearing we believe experience will demonstrate that a constant, abundant, though always fresh, supply of food ever within reach of the poultry to eat at will, is the best plan ; and we do not think the food consumed will be found at all increased, as nature and instinct will generally guide the fowls to eat no more than is requisite.

Pheasant.—At the beginning of the year, barley, corn, buckwheat, and a few white peas, constitute a good food for pheasants; and, in a cold spring, caraway, hemp, and other warming seeds may be added in small quantities. Pheasants will clear the ground of insects and small reptiles; greedily devouring even toads when not too large: they will spoil all wall trees within their reach, pecking off buds, blossoms, and leaves. The berries and insects underwood affords, are most capital food for them, as they seem to thrive on nothing so well as corn, wild seeds, berries, and insects.

Peafowl.—This bird is graminivorous, and may be fed much as other poultry, but it seems to prefer barley: its *natural* tastes in feeding, however, are almost exactly similar to those of the pheasant.

Turkey.—The food recommended for fowls generally will do very well for the Turkey, with a greater allowance of, perhaps, vegetables, which they most greedily devour. They are great eaters, and delight much in runs over fields, pasture and meadow lands, woody plantations, &c., whence they derive an incredible supply of their sustenance. Maize or Indian corn, it is said, they are very fond of.

Guinea-fowl. — This fowl requires no special plan of feeding, nor any particular description of food, although buckwheat or brank may be introduced into their diet with great advantage.

Swan.—The principal portion of the food of the Swan consists of the vegetable and minute animal matter that abounds in ponds and places of still water, and from the reeds, rushes, grasses, small fry of fish, &c., of the larger lakes and streams. When domesticated they will feed readily on the same diet as the goose, eagerly receiving grain, bread, grass, or any refuse vegetable, from the hand of their keeper.

Goose.—Although geese seem to thrive never so well as when turned upon a common or a stubble field, to seek their daily provision; a moderate supply of any kind of corn or pulse at hand must always be dispensed morning and evening to the store flocks, on their going out and returning, in the evening especially; together with any greens (which constitute an important part of the food of store geese) that may chance to be at command; as also cabbage, lettuce, marigold leaves, lucern, tares, and occasionally, sliced carrots and turnips. In some parts of Norfolk a piece of housewifery prevails, which is perhaps peculiar to that county, of giving geese, but more especially when young, feeds of *green* wheat, whenever practicable. They will rarely touch any kind of offal or animal food, as they may be said to be exclusively herb and grain feeders.

Duck.—The duck will eat flesh and garbage of any kind, yet water insects, weeds, vegetables, corn, and pulse, are its general food. But ducks seem to desire nothing so much as animal and vegetable refuse of any kind, nothing being too nasty, filthy, or putrid in that line: but we should strongly advise that, if intended for the table, they be not permitted such foul offal, nor to gorge on chandler's greaves, which some recommend, having, perhaps, better ideas of economical feeding than they have of delicate taste. Ducks will devour frogs with avidity.

Pigeon.—This bird is entirely graminivorous; eagerly eating corn and other grain, and seeds of almost any kind: but the best and cheapest food are grey peas, and the smallest sort of horse beans, (commonly called "pigeon beans,") and in this particular a dangerous mistake is often made in publications on the subject, by the writers recommending "small *tick* beans," or *kidwell* in western phraseology, which are the larger of the two common field varieties, and of inferior quality, besides being, from their large size, apt to choke pigeons feeding on them: tares are also much liked by pigeons, but there is often a difficulty in getting them in some parts. Wheat, barley, and buckwheat may also be given occasionally as a variation, though they do not seem to relish barley much: the pulse should always be old, not less than of the previous year's crop, as new is scouring in its tendency. Hemp, rape, and canary seed, though much too stimulating to be used as a general diet, are much liked by pigeons, the two last named, according to our observation, even more so than the first, which, however, is generally given by Fanciers for promoting fecundity. They will eat green vegetables sometimes,—lettuce leaves being generally selected. They also much affect salt, which may, perhaps, like the sand and lime they eat, assist digestion; but the experience gained by an experiment of a twelvemonths' duration, convinces us that it is by no means *necessary* to preserve the *health* of pigeons, as is often said, the birds in our cote thriving remarkably well during the whole of that period, without a single grain of salt being allowed them; but a pinch or two of salt thrown down occasionally on the sand or gravel at the bottom of the cote, can certainly do no harm, whilst it will assuredly delight the pigeons. Small bits of old lime or mortar are most salutary, and would seem to form an indispensable part of the food of pigeons.

GENERAL MANAGEMENT.—In addition to the directions already given, some suggest themselves, which could not properly be considered under the heads in which the subject has been treated.

Although in *naturally damp* situations, where it is necessary to *make a ground* for a poultry-yard, we have recommended a sort of concrete, of chalk, rubbish, and gravel, to be rammed down; we by no means advise that plan to be followed under other and favourable circumstances, and certainly

not that the *surface* should consist of that material,
—for it speedily becomes so hard and unyielding
as to injure fowls kept upon it, either by retaining
wet and moisture, unable to percolate through, or
by the roughness to the feet, often entailing leg
diseases : as a surface-covering for yards, nothing
is better than beach-sand, mixed with fine old
rubbish from buildings, and garden-mould. When
sand cannot be procured as the basis, gravel or
ashes may be used,—the latter being in our opinion
preferable to the gravel.

Whether the exterior roof of the poultry-house
be tiled, thatched, or boarded, the interior ought
assuredly to be ceiled; and the walls plastered.

In keeping poultry of any description, the first
consideration, after providing them with warm
and yet airy shelters, should be, that they are well
attended to, and the *strictest possible cleanliness*
observed in their houses; otherwise they will soon
be, in the vulgar phrase, " stenched out," or in-
fected with the impurities they occasion in the at-
mosphere. The floor of the house must often
(daily, if possible) be swept down, and fresh strewed with sand, ashes, or gravel,—saw-dust, recommended by some, we think injurious to fowls.
The ceiling and walls must be white-washed with
lime three or four times in the year—the warmer
months rendering that operation more especially
necessary. The laying and hatching nests should
also be lime-washed frequently—the latter after

every brood has been raised. This precaution destroys lice, if such exist; but if any difficulty is experienced in getting rid of those disagreeable lodgers, smearing with turpentine, or fumigating with sulphur on a pan of charcoal, will seldom fail. The surface of the yard itself, being raked over and cleansed thoroughly every week at longest, may be allowed to stand three or four months, when, for some two or three inches, it should be removed, and fresh covered. To carry these instructions out as they require to be, it must be obvious that, if many fowls are kept, unless the proprietor is prepared to devote an amount of time, care, and attention, which very few we think will, it will be necessary to have an attendant for the yard: if not very extensive, a shrewd, careful, and *humane* girl is, perhaps, the best adapted of any to take charge of poultry,—at any rate, whether male or female, the poultry attendant must, for the well-doing of the stock, be kind, gentle, cleanly, and untiring; but no provision as to attendants will relieve the owner of the fowls of a responsibility peculiarly attaching to him,—that of now and then casting a searching and superintending eye upon the stock, houses, yards, food, &c., from which incalculable good will result.

It is generally recommended to leave the wicket-gates at the bottom of the doors of the poultry-houses open at night, in favourable weather at least, in order to afford the fowls an opportunity of going out abroad at early dawn; but this we think bad in many ways: rats or other vermin may thereby get in during the night,—and, which will even prove more extensively destructive of the stock, (especially when young,) the fowls will sustain great injury in their health, from venturing over the grass or ground when loaded with heavy dew. If it is desirable to let them out so very early in the morning, the only proper way is for some one to be up to let them out, if and when he knows they will not take harm.

Pullets and cockerels should be separated when about nine or ten weeks old,—and on no consideration permitted to range together after they have attained three months, as they will breed too early both for their own health and that of the progeny.

The food and water of all poultry should be given in such a way, as to preserve it from contamination, as much as possible, arising from dirt, &c. To this end pans or troughs are now generally provided; and convenient ones, of iron, may be obtained of the ironmongers, at very reasonable prices: these are constructed in various forms,—flat like a tray, circular, semi-circular, or triangular for standing in corners; and have one or more compartments, to hold water, grain, or other food. The best, but most expensive, are the double troughs, with bars across to prevent wasting or soiling the contents, and so arranged, that whilst they may be lifted off for cleansing, yet

2 H

fowls themselves cannot rest upon them. If the breeder adopts the plan of feeding his fowls by hand only, or has the food contained by itself in a pan, or wooden hopper, the water may be kept either in the common earthen fountain bottle; or those to whom expense is not a consideration, can procure one of Baily's registered fountains, which are certainly better, and much easier to fill and cleanse.

Laying hens are great devourers of lime and chalk in any shape,—and, indeed, it seems almost necessary for the health of the fowls. Care should, therefore, be taken that plenty of old mortar, lime-rubbish, chalky marl, crushed bones, and oyster-shells, be always strewed about the yards and houses.

Of course, if it is intended to confine swans to any particular spot, it seems superfluous to state, that if the stream or piece of water on which they are kept, has any outlet, it will be necessary to close it for some yards up with netting, or by other means. Some breeders place a feeding-trough by the bank-side, to keep the stock supplied when there is paucity of food in the water; but a far preferable plan, and more natural, is to cast a handful or two to them on the water. At two or three points the bank must be made sloping, so as to form what may be called landing-places for the cygnets or young birds. To retain swans, the enclosure of the stream will not alone suffice, if left the free use of their wings: it is usual, therefore, to *pinion* them,—an operation which is performed either at the first (or wrist) or at the second (or elbow) joint of the wing; and we think the second will be found necessary for the effectually securing swans. It must be done cleverly, with a very sharp knife, just at the socket of the joint: any bungling or mangling will endanger the life of the bird. Cygnets intended for fattening for the table should not be pinioned, as the loss of blood weakens them, and has a tendency, of course, to retard the fattening process. From the tyrannical conduct exhibited by them towards other fowls, which they will scarcely tolerate near their domains, if other water-fowl are kept on the same stream, it is customary with many to slit the webs of the feet of the swans, as otherwise they generally put to flight and disperse every species but their own,—and even these will not always be spared. It is not popularly known that swans thrive better on shallow swampy waters than on deep running streams. In confinement, from frequent diving and explorations into the muddy beds of their waters, swans frequently have an accumulation of mud-deposit under the lower bill, which the attendant must occasionally remove by the hand, to free them from the dirt. The reputed longevity of the swan exceeds that of any other animal; and it would form an interesting record, if present and succeeding proprietors of swanneries would keep a stud-book of their stock, with the ages carefully noted down, and identified by certain marks, in order that naturalists might be better and more accurately informed on the point.

In commencing to stock a dove-cote, it is always preferable to purchase *squeakers*, or young ones which have never been flown from the home in which they were reared; as these, being confined, in a short time, well fed, and accustomed gradually to the surrounding country, before they have attained sufficient strength of wing wherewith to lose themselves, will become perfectly domesticated. Indeed, the difficulties of accustoming or keeping *old* pigeons to a new home, are almost insuperable, and without the greatest vigilance, or cutting the wings often, they will speedily take themselves back to their former abode. The management of the dove-cote requires the greatest possible cleanliness to be observed, as nothing injures pigeons so much as a foul loft, which should therefore be well cleaned out and ventilated every day; also lime-washed very frequently, as apart from the purifying effects produced, the pigeons exhibit an unmistakeable preference for the white colour, which seems peculiarly grateful to their eyes. Fine sand and gravel should every now and then be strewed over the bottom of their loft, with a supply of bricklayer's lime-rubbish always at hand, for the quantity they will eat is enormous.

Pigeons are exceedingly fond of bathing in clean water, and every day a pan of fresh water should be placed in each division, when they will be seen to bathe and cool themselves in it by the hours together. The drinking water is kept cleaner and fresher in one of the iron or earthenware fountain bottles already recommended for the poultry-yard; but even this reservoir cannot be too frequently washed and re-filled. The ordinary food is generally given in a *meat-box*, which is formed in the shape of a hopper, covered at the top with a sloping roof, to keep clean the grain, which descends into a square shallow box at the bottom. Some fence this with rails or holes on each side, to keep the grains from being scattered over; others leave it quite open, that the young pigeons may the more easily find their food. When any small dainty, as hemp, rape, or canary seed is given them, pigeons seem to prefer to search for and peck it from among the sand and gravel, whence it is as well to throw it at the bottom of the cote or loft. The strong scent of cummin, and flavour of coriander seeds, are said to have an alluring effect upon these birds; as also the scent of *assafœtida*, and other powerfully odoriferous drugs; and that the use of such will attach the pigeons to their home. The last article necessary to be described, is the *salt-cat*, so called from some old fancy of baking a real cat with spices, for the use of pigeons. We have placed in the middle of the loft a dish of the following composition: loam, sand, old mortar, fresh lime, bay-salt, cummin, coriander, caraway-seed, and all-spice, moistened into a consistence. The pigeons are frequently pecking at this, and are in a constant state of health—how much of which may be attributed to the use of the cat, we cannot determine; but, certainly, they are extremely fond of it, and, if it has no other merit, it prevents them from pecking the mortar from the roof of the house, to which otherwise they are much inclined. In the attendance at the loft watchful caution is necessary to prevent them from, or separate them when, fighting—an amusement they are much more prone to than might be expected, often to their own serious injury, and the destruction of their eggs or young.

TREATISE

ON

THE DISEASES OF POULTRY,

BY

F. R. HORNER, ESQ., M. D.

WITH regard to the *Diseases* of poultry, feeling our own incompetency to treat the subject as it should be, we preferred candidly to admit it, rather than add to the nonsense already written upon it. We therefore applied to F. R. Horner, Esq., M. D., (to whom we are indebted for many other valuable hints and suggestions,) for his assistance in elucidating that department: and upon our earnest solicitation, and at considerable sacrifice to his own feelings, that gentleman at length most courteously consented to put his ideas upon paper; and the result has been the production of the following highly valuable and original Essay, in which the whole matter is so scientifically and yet popularly treated, that we trust the learned Doctor will feel rewarded for permitting its publication, by the reflection that he has done good service to the cause of science,—as, indeed, it is only thus we can hope to disseminate truth, and disperse accumulated error and ignorant quackery: for it must be obvious, that the subject can only be properly treated by some one of extensive observation and experience, and who is, above all, acquainted with the *anatomy* and *functions* of fowls,—and the justice of the preliminary remarks on this head, by Dr. Horner, confirmed as they have been by some extraordinary and interesting experiments, must be at once apparent to the reader.

THE treatment of the diseases of poultry has ever been chiefly empirical; and at the present day, nostrums and specifics to cure all their disorders, are continually being set forth. Moubray has observed, with much severity and truth, that "the far greater part of that grave and plausible account of diseases and remedies, which is to be found in our common cattle and poultry books, is a farrago of sheer absurdity; the chief ground of which, it is to be appre-hended, is random and ignorant guess-work." The attempt has, however, been made by some recent writers, to describe the diseases of fowls more scientifically: and doubtless, they have rendered very great service; yet, it must be conceded, that sometimes, in their advocacy of new views and the ap-plication of some remedies, their efforts do but teach us how especially watchful we ought to be, that, in treating of the diseases of poultry scientifically, we do not, at the same time, do so irrationally.

The great source of error arises from disregard of the im-portant fact, that there is a vast and intrinsic difference be-tween the constitution of fowls and that of man;—not only in the structural or anatomical, but also in the functional conditions of the great organs of their bodies, and especi-ally in those of digestion and of respiration, as well as of the skin. Indeed, the function and condition of the skin of fowls, bears no analogy to that of man. In him, it not only exercises most important influence in health, but in disease it oft is the great medium by which medicines effect relief. Hence it follows, that the action of a certain class of medi-cines on fowls must be negative. Yet we read, that cer-tain drugs are recommended and administered to fowls, which, if there be any truth in the position just advanced, cannot exert upon them that action or effect, which they do on man, when afflicted with similar disease of the same parts.

It is not enough that we possess a knowledge of the medicinal powers of a drug, when applied to the diseases of man; nor even, that we are acquainted with the real nature of the disease with which a fowl is afflicted: the important consideration still remains, how far the anatomical and functional peculiarities of parts or organs, will modify or change, will increase or (may be) wholly negative, the action of a medicine on their so differently constituted bodies.

Let us take an example. The tartrate of antimony (tartar emetic) has, by recent writers, been much commended in some inflammatory diseases of fowls,—as in inflammation of the windpipe, and also of the lungs. It is well known, however, that much of this drug's beneficial effect in those diseases of man, is attributable to its action on the skin—to its inducing perspiration, or promoting expectoration; and thereby affording relief to the oppressed internal parts. As the skin of fowls, however, covered with feathers, does not perspire,—as fowls do not expectorate, as man does,—reason tells us, that there can be no identity or true relation in the curative action of the medicine. We do but witness the anomaly of a drug being administered, that cannot exert on fowls those very influences, on which, in man, its efficacy so greatly depends.

Again, when we consider what may be termed the triple organs of digestion in fowls—the crop, the fore-stomach, and the gizzard; or, in plain explanatory language, the steeping vats and the mill; organs, possessed of such power, that the most heterogeneous substances, as hard grain, seeds, grass, and vegetables, worms, insects, &c. &c., are readily reduced and converted thereby into one homogeneous mass of food; we may easily conceive, that some drugs also may undergo in the stomachs of fowls such amalgamation, trituration, and essential change, as greatly to modify or obtund their medicinal effects. Repeated experiment has, indeed, shown to me, that such is truly the case; the results being, in some instances, as extraordinary as they ought to be instructive. A few of the most remarkable only will, in the briefest manner, be alluded to; and those drugs selected, whose action on the human body is constant and well ascertained.

Five-grain doses of Turkey opium (more than a quarter of an ounce of laudanum) had no effect whatever on Polish hens. Eight grains (half an ounce of laudanum) were then given to Cochin China hens, without any effect. Lastly, sixteen-grain doses (equal to one ounce of laudanum) were given to each of two Cochin China hens; the one which had a young brood of chickens, was observed to brood them for a longer interval than usual, and appeared somewhat dull for four hours, but she at once made off with her chickens upon being approached; the other was not perceptibly affected, and both took food as usual. The Cochin China with chickens was observed to be briskly purged by the dose; in the other, there was no difference in this respect; and in none of the cases were the bowels at all constipated. An ounce of laudanum would certainly be a poisonous dose for a man; in many instances, half an ounce would be the same. Hens are not the only creatures that are so insensible to the effects of opium; a rabbit, physiologists tell us, will take as much with impunity as would prove fatal to a dog.

Twenty grains of ipecacuanha (an emetic dose for a man) were given to Polish hens; they ate as usual throughout the day, and were lively; the bowels were slightly purged—no other effect whatever was witnessed.

Twenty grains of catechu, a powerful astringent, and the full usual dose, had no effect on the bowels of fowls.

Tartar emetic in two-grain doses had no effect on Polish hens, the birds taking their food as usual. (One grain is the usual emetic dose for man.) In five-grain doses it affected the bowels, and food was refused. In ten-grain doses it acted as a brisk purgative, and all food was refused for a few hours. The same large doses of ten grains were then given to a Game and to a Cochin China hen, which had just been purposely fed with barley; the digestion of the food was arrested, and both the fowls were seriously affected with spasm, the head in one case being drawn to one side: they were with difficulty recovered. These two examples are instructive; showing, that the crop, being full of grain, the irritant effects of the tartar emetic could not then be negatived by the secretions and actions of the digestive organs. Ten grains of tartar emetic would certainly produce inflammation, and probably death, in man, if it were not previously rejected by vomiting.

Five grains of arsenic, and five grains of corrosive sublimate of mercury, were severally given to Polish hens without any effect; except that, like the tartar emetic, these most poisonous minerals passed off innocuously by the bowels. The sixteenth part of a grain is the usual dose for man.

Calomel, even in doses of twenty grains, does not act severely on the bowels of poultry; but in doses of four or five grains its aperient action is discernible.

These experiments are suggestive of many practical points; and show, 1st, that, if such powerful drugs have so little or even no effect on poultry, even in doses which are poisonous to man, it is highly questionable whether they can be beneficially employed at all, in the cure of their diseases. 2nd, that, in the comparatively minute doses in which some of them have been frequently recommended by recent writers, they cannot, in reality, exert that curative agency which has, unwittingly, been attributed to them. 3rd, that the mere size of the creature, as compared with man, is no guidance for proportioning the dose of a medicine which will affect it.

Thus it appears, that, not only is the digestive process of fowls so essentially different to that of man; and hence the effects and changes produced on matters introduced into their stomachs, whether food or medicine, also different; but, what is of the highest moment, other organs and structures of the body in fowls, bear little analogy or resemblance in their functions to those of man. Hence, it is wholly irrational to conclude that, when poultry fall into disease, they can be cured by medicines which, in man, produce their curative results, chiefly by their influence on associated and, may be, distant organs.

APOPLEXY.—This disease occasionally occurs amongst fowls, and is necessarily of a most dangerous character. It is by no means restricted to the aviary or poultry establishment of the amateur; being, perhaps, equally of occasional occurrence in the farm-yard; but, being there less cared for, or understood, is passed over. It does not appear, that one variety of fowls is more subject to it than another: chickens, or young fowls, are less frequently attacked than older ones.

Symptoms.—In some examples, the bird suddenly falls down, and after a few convulsive motions, dies. It most frequently happens, that it is affected with "staggers," for one or more days before death; or the head is gagged back upon the neck, or twisted forcibly to one side. Moving about being difficult, it is commonly found in a sitting posture, or half turned on the back. If now disturbed, its gait is staggering, and it seems involuntarily impelled to go always to one side: superadded to this, attacks of convulsion occur at intervals; in one of which the poor bird at last expires.

Cause.—The appearances after death are, either effusion of blood on the brain, or, otherwise, a turgid and congested state of its blood vessels. Although it is usually supposed that this disease is the effect of a system of high feeding, yet statistical inquiry does not appear to support the opinion. In the farm-yard, and in Zoological gardens, where birds are not over-fed; and also, in those establishments of amateurs where the poultry are fed but moderately, apoplexy is ever of occasional occurrence.

Treatment.—When man is the subject of apoplexy, the most discriminating judgment of the physician is, in some cases, required, to determine whether bleeding, or a very opposite treatment, should be immediately adopted; the life of the patient being, as it were, in the balance, and dependent upon the accuracy of his decision. The poultry amateur, however, has few opposing data, or diversity of symptoms, either to perplex or to enlighten him: and upon the whole, therefore, he will do well in all cases to bleed the fowl threatened or attacked with the disease. The vein running near the bone, at the under side of the wing, is the one recommended to be opened: if relief be not afforded, the bleeding should not be repeated. In all cases, a brisk aperient, which may consist of 20 grains of jalap with 5 of calomel, should be at once administered. Aperients, indeed, daily, or every other day, will now comprise the best medical treatment. The fowl should be kept quiet, and have a moderate allowance of soft food.

COMB (WHITE).—The title sufficiently indicates the appearance of this complaint.

Symptoms.—It usually comes on as a few white, mouldy-like points, which gradually unite and spread; not only over the comb, so that it becomes wholly white, as if dredged with flour; but the scalp and upper part of the neck are, in many cases, denuded of feathers, when the skin presents a furfuraceous or branny appearance. Cochin China fowls seem to be especially liable to it, though it also attacks other varieties.

Causes.—Cutaneous diseases are frequently but symptomatic of constitutional disorder, and especially of primary derangement of the digestive system; as a natural sequence of which, "the life" of the creature—the blood—is oft disturbed in the relative proportion, or quality, of its component parts; and which is evinced in various forms of external or internal disease. The skin is of vast and varied extent; inasmuch as, the mucous membrane which lines the mouth, the throat, the stomach, and bowels, as well as the lungs, and all parts or cavities which have an external aperture, is but a continuation of the skin which covers the external body: the reader may readily conceive, therefore, how the disordered state of one portion, as of the stomach, may even directly and at once excite disorder in another part, as in the external skin; whilst, on the other hand, how many internal diseases are greatly relieved by medicines, which can excite perspiration, or the natural function, of the skin. In man, how often will a few pickles, a crab, or a lobster, by irritating the internal skin, or lining of the stomach, cause the external skin to be immediately covered with nettle-rash. Want of cleanliness, confinement, and unwholesome diet, are also exciting causes of this complaint.

Treatment.—We are taught by experience that, whether the exciting cause of skin diseases be internal or external, in many cases, they are capable of being arrested by external applications. White-Comb is of this class; and it readily disappears by the use of local applications. I have found, that a linament, composed of equal parts of lime water and linseed oil, applied with a feather night and morning, soon removes it. Its great efficacy in many human cutaneous diseases, happily suggested its trial in White-Comb. Should an inveterate case present itself, the addition of one-third the quantity of citrine ointment, would increase its activity: this, however, has never been found necessary. Cocoa-nut oil and turmeric, in the proportions of the quarter of an ounce of the former and half an ounce of the latter, are stated, on undoubted testimony, to remove the diseased state of the parts. Of course, the remedies should be used to the denuded parts, as well as to the comb itself.

Internal medicines should, however, be also given, that the return of the disease may be more certainly prevented. An occasional aperient of 20 or 25 grains of jalap, and also 5 grains of mercury with chalk, (the *Hydr. cum Creta,*) every night for a week, ought to be administered.

CRAMP.—This affection of the limbs is so apparent, and well understood, that it appears unnecessary to dwell upon it at any length.

Symptoms.—Great difficulty, or total incapability, of walking or standing, with a contracted state of the toes and of the limbs, marks the disorder.

Cause.—Damp and cold, and especially dampness of the ground, is the especial cause.

Treatment.—Warmth is the great restorative; even the warmth of a fire should be resorted to, in the case of chickens; and the sooner the better. The experienced amateur will take care, that the hen be not allowed to take her very young brood amongst wet grass, &c., after rain; a precaution too often neglected, indeed; hence, not only cramp, but early death, is the result. The sportsman well knows how fatal to his hopes is a wet early summer—early enough fore-shadowed to him, as he views on the foot-path of the meadow, the cramp-stricken form of the young partridge.

CROP-BOUND.—This occurrence is sometimes witness-ed; attention being usually drawn to the fowl, by its dull and moping habits; when, although the crop is distended, it is observed to attempt to eat and drink at short intervals; for, as not any food now passes into the stomach, it suffers from the cravings of hunger.

Symptoms.—These are, simply, a permanently distended crop, with a dull and depressed appearance of the bird, and frequent attempts at eating and drinking.

Cause.—In a few instances, the distention is caused by the inlet to the fore-stomach, from the crop, being occluded by some large object, as a bone lying across it. In most cases, the cause is not apparent; we simply find the crop distended with grain, which, from its long detention, is in a swollen and, in some portions, semi-putrid state. There is evidently a temporary loss of power in the propelling action of the crop. It is usual to attribute this to over-distention in feed-ing, but it is highly probable that other circumstances lead to this condition—such as, the impaction of two or more corns of barley, or other grain, in the passage from the crop; or disorder of the functions of the crop itself, inducing a partially debilitated state; such condition would, no doubt, favour distention: these are but surmises, yet probable ones.

Treatment.—Happily, the remedy is always at hand, and is invariably successful; viz. incision of the crop and empty-ing of the contents. The part at which the crop should be opened, and the size of the opening, are very important; in order that ligatures, or sutures, may be unnecessary. With a lancet, sharp scissors, or pen-knife, an incision, little more than a barley-corn in length, should be made at the upper

part of the crop; the swollen and impacted contents must
then be carefully dislodged. A bodkin, or large needle, or
the blade of a pair of scissors, or similar domestic instrument,
which is always at hand, will be quite efficient for the purpose. The crop being then emptied, the bird may be allowed a *little* water from time to time, as there is much
thirst. The food must be restricted to very small quantities
at a time, and should be of a soft quality, as bread-crumbs,
boiled rice, or barley-meal : if food and drink be not so restricted, the fowl will partake much too freely, and, by distending the crop, keep the wound open. Though the parts
are so very thin, they heal with great rapidity, if left, undisturbed, to nature. The bird does not evince any suffering
under the operation; indeed, care must be taken that
it do not again eat the grain, as it is being dislodged, and
which hunger always impels it to do. It must be cooped
for a few days, but only that the quantity of food may be
more certainly regulated.

CROUP, or *Tracheitis.*—Fowls, when young, are very
subject to inflammation of the inner or mucous membrane
which lines the wind-pipe (or trachea). It is most commonly of a sub-acute or chronic character, and occurs more
frequently in some seasons, and also in some localities, than
in others. Many chickens, or young birds, are carried off
by this disease; sometimes after a short illness, but more
generally they linger for some weeks.

Symptoms.—Difficult breathing is the most distinguishing
and prominent symptom; and it is highly important, in
enabling us to form an accurate judgment of the nature of
the disease, and to distinguish it from inflammation within
the chest, that the kind or character of the breathing be
especially noted. When the lungs themselves are inflamed,
the respiration is short, panting, and quick; whereas, in
Croup, it is slower, oppressed, and generally noisy; the fowl
always breathes through the mouth, the bill remaining agape;
or otherwise, it is opened at each respiration, with a slight
noisy click. The bill is also somewhat elevated, by the

head being thrown back, to facilitate the entry of air through the windpipe, by straightening the neck. A glutinous, mucous secretion, for the most part, appears within the mouth or throat, which the distressed bird occasionally makes efforts to dislodge. When this mucous matter becomes dried, and hardened on the tongue, by the passage of the breath over it, it was formerly called the *Pip;* but it is evidently nothing more than a symptom; and one which may arise in other diseases, where the breathing is performed through the mouth, rather than through the natural passage, the nostrils. It appears probable, that the term *Pip* has been given, as being expressive of the peculiar noise made by the fowl, in its efforts to remove, by expulsion, the secretion from the throat or mouth; it is something between an apparent act of sneezing and coughing, and is accompanied by a quick jerking motion of the head.

On examination after death, the mucous or lining membrane of the windpipe is found thickened; and there is adhering to it a plastic or viscid secretion, of a yellowish white colour, the result of inflammation. In some chronic cases, it is observed, that this adhesive secreted matter is found, though in less quantity, in the two larger air tubes, where they unite with the windpipe; showing extension of the disease. A decided "false membrane," as it is called, of the tubular form of the windpipe, has not been observed by me in fowls. Such "casts" of the tube, are seen in the *acute* Croup of children only : in chickens, as before stated, the disease is rather of a *chronic* or *sub-acute* character.

It is worthy of remark that, whether Croup occurs in mankind or in poultry, it is the young which are attacked.

Causes.—Cold easterly winds, wet weather, and especially damp, foul, and cold roosting-houses, are the great exciting causes of this inflammatory disease.

Treatment.—In this, as in most other cases, it is by attention to regimen, rather than by giving medicines, that we shall restore the sick fowl: at all events, removal to a warm, dry, and clean situation is the first and most important step. Moubray strongly insists on the great benefit of

2 i

warmth, he says, " the sun, or warmth in the house by the fire-side, are the best remedies."   Diet must also be attended to; and should consist of the milder kinds of food, such as rice, barley-meal, mashed potatoes, bread, separately or variously combined, and such like.   I must especially insist upon the importance of *green food* in diseases like this—chickweed and lettuce are the most beneficial, and by far the most relished; they should be cut into small pieces, and offered alone, and if not freely partaken of in this way, they must be mixed up with the soft food.   Lettuce also contains a considerable portion of a sedative or narcotic principle, *Lactucarium*, which is beneficial in irritation of the lungs in man.

The mild dose of an aperient, (according to the age of the bird,) as five grains of jalap with one of calomel, must be given at the commencement; and about five grains of the *Hydr. cum Creta* (mercury with chalk) may be administered every other night.   Should the disease continue for about three weeks, the cayenne capsules (see *Roup*) ought to be given night and morning: from two to four (or more) as the age of the fowl may be, crumbled small with the fingers, and made into a pill with butter, is the best way of giving them.   It is certain that this remedy possesses great influence in some *chronic* diseases of the mucous membrane of poultry.

The treatment then is comprised in these regulations : a warm, or rather an artificially warmed, abode; mild and soft food, with chickweed or lettuce, and pure water; an occasional aperient, and a few grains of mercury with chalk, every other night; and when the disease is prolonged, and the debility great, the cayenne capsules twice a day.   The *Hydrargyrum cum Creta* (mercury with chalk) has an alterative and beneficial effect, in the sub-acute Croup of young poultry.

DIARRHŒA.—This sometimes proves a very weakening disorder, and one which is apt to be too much neglected.   In chickens, or young poultry, its continuance is severely felt, and induces great prostration. If it do not abate of itself, in 36 hours, remedies should be had recourse to.

Symptoms.—These are sufficiently obvious: it is important, however, to notice the manner in which the evacuations occur. If they are accompanied, or immediately followed, by a jerking, or suddenly retracting motion of the body, we may conclude that the lower bowel is suffering great irritation; and if not relieved, may terminate in dysentery, or inflammation.

Causes.—Exposure to cold and damp, either in the roosting-house or elsewhere, will induce the complaint; and that in its most severe form. It often arises also from errors in diet, as sour, unwholesome, and *want of variety* of food, and especially from the too exclusive use of that which is of a soft nature.

Treatment.—If induced by cold and wet, it will be especially required to keep the birds in a warm and comfortable place for a few days; and to supply them with mild nutritious food, such as rice, barley or oat meal, &c., dredged occasionally and moderately with black pepper. If connected with unsuitable diet, this must be at once corrected.

Aromatics, and especially when combined with chalk, form the most suitable and efficient medicines. Twenty grains of the compound chalk powder (the *Pulv. Cretæ comp.* of the Pharmacopeia) night and morning, as a bolus, will, for the most part, be sufficient.

EGG-EATING HENS.—This untoward propensity has been attributed to various, and even opposite, causes. Some consider it as a mere bad habit, induced by the practice of giving egg-shells; whilst the lack of calcareous matters, as lime, egg-shells, mortar, &c., has, on the other hand, been supposed to have given rise to it. Be this as it may, it is certain that the abundant supply of these substances has neither prevented nor corrected the habit; nor has personal correction itself with switches been more generally successful.

It appears highly probable that this unnatural propensity

is connected with a vitiated or disordered state of the digestive system itself; and is somewhat analogous to a similar false appetite, or "longing," not unfrequently occurring in the practice of the physician in our own species; namely, the uncontrollable desire for eating such things as dry or uncooked rice, dry coffee grounds or berries, chalk, slate pencil, and even cinders, &c. Such symptoms are removed without difficulty by appropriate remedies: and in the case of egg-eating hens, the treatment recommended is, moderate aperient doses of jalap with cayenne capsules, ten grains of the former and three of the latter, every other day; and a grain of calomel every night for a week. These alterative and aperient medicines constitute the most rational and effective treatment.

GAPES. (See *Parasites.*)

GOUT.—Although it has a place in the long list of ills that poultry is heir to, it is here disallowed; as it is considered, that its existence in fowls rests but on fanciful or mistaken evidence.

LAMENESS.—Under this head, a not unfrequent cause of lameness, called the bumble foot, may cursorily be mentioned. It is rarely seen except in heavy-bodied birds, as the Dorking.

Symptoms.—Although it at first commences as a circumscribed swelling, or small tumour, at the under part of the foot, it at length extends, and involves the root of the toes; so that the unsightly size of the parts has gained for it the appellation of bumble-foot. The swollen and inflamed part sometimes eventually ulcerates, so that the poor bird is quite lame.

Causes.—Narrow and high perches are evidently the chief origin of the disease. They both cause injurious pressure to be made on the feet during roosting, and also great concussion when the unwieldy bird flies from them on to the ground.

Treatment.—If the disease be discovered early, its progress may be stayed by the substitution of broad and low perches; or, what is better, by making the injured birds to rest, for some months, on the ground, strewed with a little straw. Happily, high and narrow perches are now, nearly everywhere, discarded from poultry-houses.

LICE.—These become very annoying, and when, through want of dust-baths, they accumulate in excessive numbers, are prejudicial to the health of fowls. A dusting-hole in the ground, (which is always preferred,) or a box containing pulverized soil, finely powdered old lime rubbish, or, what is very efficacious, cinder dust from the hearth, should always be considered an indispensable adjunct to the poultry-yard. Flour of brimstone, either dusted upon the skin, or mixed with the contents of the dust-bath, is highly recommended by both old and recent writers. Mercurial ointment, a little being placed on the skin under the wings, also quickly removes these parasites.

LUNGS, INFLAMMATION OF.—Bronchitis, or inflammation of the lining membrane of the air-tubes of the lungs, is not an unfrequent disease, in the cold or winter months of the year. It occurs amongst fowls of all ages; but the old and the young are most frequently affected.

Symptoms.—The most prominent symptom is, a noisy and laborious respiration, impressing the observer with the fact, that the air tubes are loaded with phlegm. Though the breathing is thus rendered difficult, the bill is not habitually open, as in Croup; the oppression being rather in the chest than in the narrowed air-tube,—the wind-pipe, or the glottis. Thirst, though present, is not extreme: there is an appearance of great dulness and oppression, as well as loss of strength. The complaint generally appears in a sub-acute or chronic form: some fowls, and more especially aged ones, are subject to it yearly, on the approach of winter.

Causes.—Cold, wet, and variable weather, is the common cause.

Treatment.—All those regulations in respect to warmth and dryness, as also diet, which were enforced in the treatment of Croup, must be duly observed in Bronchitis; the same general plan of medical treatment is also applicable.

PNEUMONIA.—This disease differs from the one just treated of, in this respect, viz. that it is the substance of the lung itself, and not simply the lining mucous membrane of the air-tubes, which is the seat of the inflammation. It is less frequent than Bronchitis, and occurs rather in the acute than in the chronic form.

Symptoms.—The character of the breathing differs much from that of Bronchitis. In Pneumonia, it is short and panting; and not laborious, prolonged, and accompanied with a rattling in the chest or throat. The thirst, too, is very urgent; and while the bird seeks to be alone, the feathers are generally much ruffled, erected apart, or puffed up; there is also much fever present.

Causes.—The great exciting cause is also cold.

Treatment.—All those means of warmth and comfort, already so frequently pointed out, must be strictly adopted; the diet must be mild and unstimulating; and of which green food must form a liberal portion. Pneumonia and Apoplexy are the only two recognisable diseases of poultry in which bleeding is admissible: the operation should be performed as directed when treating of the latter disease. A purgative of jalap, say twenty grains, or ten grains with five grains of calomel, must be given immediately, and should be repeated two or three times, at intervals of two or three days. It is also necessary to give a grain of calomel, or five grains of mercury with chalk, (*Hydr. cum Creta,*) every night. A little nitre (saltpetre) dissolved in the water for drinking, as recommended by Moubray in Roup, is in accordance with just principles of treatment, and appears serviceable in the Pneumonia of fowls, as it unquestionably is in that of man.

MOULTING.—Though this is a natural, and not a diseased process, yet it is a trying period for fowls that are out of health, and also for the aged; hence it is by no means an unfrequent thing for them to die at this time. The difficulty of moulting increases with age, so that not only do aged birds moult later in the season, but the duration of the process is protracted.

Treatment.—The great point to be attended to, and especially when the fowl is weakly, either from age or disorder, is warmth—it is all-important; and if any danger be apprehended, a dry and warm habitation should be provided: all birds, indeed, moult more rapidly and kindly when warmth and dryness are secured to them. The diet. must be mild and nourishing; green food being also liberally supplied. If there be a want of appetite, and the fowls appear out of order, an occasional mild aperient dose of jalap with cayenne capsules, ten grains of the former with five of the latter, may be given; or, a pill of rue, the size of a horse-bean. Though Moubray tells us, that he does not know that this herb has any effect at all on poultry; other writers now denounce it as dangerous to them; whilst vulgar experience elevates it into a very panacea for all their ills. Certain it is, that fowls have quickly devoured a whole rue-bush in a small enclosure where there was no other green food. As elsewhere explained, in large doses it acts as a stimulant aperient; in smaller ones, simply as a grateful stimulant: in acute inflammatory diseases, it should not, of course, be given.

OVARIES (*Soft Eggs*).—The ovaries of laying hens are liable to disturbed or disordered action; when the result is, soft or shell-less eggs.

Causes.—This disordered state appears, in a few instances, to be connected with debility, direct or indirect; or on want of healthy tone and power in those organs; and then, usually occurs in hens that are kept in too great confinement, and much fed up: mere fatness in hens being no more evidence of healthy vigour than it is in the stall-fed fatted ox. It very frequently, however, depends upon excitement and irritation of those organs, produced by too stimulating food. The vulgar notion, that soft eggs are always caused by want of calcareous substances, as mortar, chalk, &c., is mere unfounded supposition; and hens are often known to lay soft eggs, when those very things are abundantly supplied.

Nature, in her vital laboratory, our bodies, forms from the natural food we eat, all those ingredients, or constituents, required for framing its various parts; thus, she makes enamel for the teeth, lime and phosphorus for the bones, iron for the blood, and the soft pultaceous substance of the brain! Of course the food must be natural, as nature wills it.

Treatment.—Regimen, which term embraces the regulation of food and water, liberty of exercise, ventilation, cleanliness, &c., is the first thing to be attended to. It is certain that the symptom often ceases on a change of diet; the mere substitution of wheat for barley has, in some cases, forthwith produced the wished-for change. Wheat, rice, potatoes with barley-meal, and such-like, comprise most suitable food; whilst, where too great excitement exists, barley in the grain, greaves, and all stimulating diet of a like quality, must be withheld. Especial care must be taken that green food be liberally supplied. As a medicine, jalap, combined with two grains of calomel, or alone, in moderately aperient doses of fifteen or twenty grains, and repeated every third day, will be found a most efficient medicine: should the hen appear weak and out of health, although fat, the jalap will be advantageously combined with a few of the cayenne capsules; or the finely-minced leaves of rue, made into a bolus with butter, the size of a nut, may be substituted; as it possesses, in small doses, also a grateful stimulant action. In such cases, *variety* of food is also wholly indispensable.

PARASITES.—It has been deemed prudent to abolish in the present treatise the old appellation of " the Gapes ;" for it is manifestly irrational, as Moubray justly complained, " to coin a new disease," which is, truly, but a symptom ; and one, too, that arises from wholly different causes. This excellent old author evidently doubts the existence of " the species of fasciola infesting the trachea or windpipe : " but the testimony of many writers assures us, that small, red, worm-like parasites, are sometimes found in the windpipe of fowls ; and which, from the irritation and partial obstruction to the breathing they induce, frequently cause the bird to gape, and to stretch or twist the neck. Doubtless, however, much fatal error in treatment has arisen from the bold assertion, that these parasites are the sole or invariable cause of the symptom. For, in Tracheitis, or Croup, like symptoms occur ; and especially when there is formed in the windpipe, as the effects of inflammation, a more abundant quantity of the plastic, mucous, or gelatinous-like shreds, before described. These, causing obstruction to the breathing, also excite the fowl not only to gape, but, as in the case of Parasites, at times to stretch or turn the neck, in its efforts of dislodgement.

Symptoms and Cause.—The symptoms have been already noticed in the foregoing observations. The cause being the presence of slender, red, worm-like parasites, which attach themselves, by a hooked tentaculum, to the mucous membrane of the windpipe.

Treatment.—It appears that they are readily dislodged, by inserting, and then twisting smartly round, in the windpipe, the top of a feather which has been dipped in spirits of turpentine. Another method is, to place the fowl in a closed box, and then exposing it to vapour of turpentine, the spirit having been sprinkled freely therein. Spirits of turpentine, it is well known, is a direct poison, even to the tape-worm itself.

It is, however, necessary to remark, that the advocates of the sole parasitic theory of " the Gapes " themselves report, that fowls are sometimes killed by, or die shortly after, the use of the turpentined feather ; and further, that the parasites are not always discoverable. It is quite apparent, indeed, that this application has often been employed, when the symptom truly depended upon inflammation of the windpipe, and the secretion of those gelatinous-like shreds afore-noticed; in which case, death must needs be the result of such measures: to guard against so unfortunate an event, a due consideration of *all* the attendant symptoms of Croup, before specified, will be the best surety.

RHEUMATISM.—There can be no doubt but that fowls are liable to a lameness, which is truly of a rheumatic character. It is most prevalent among young poultry, from their greater susceptibility to be affected by the exciting causes of this disease.

Symptoms.—Lameness, difficulty, or even incapacity of walking, is the prominent symptom, not arising from accident or injury.

Causes.—Cold and damp are the causes of Rheumatism in fowls, as they are in man: the muscles of the legs, the muscular and ligamentous parts of the back, are those generally affected; and in some instances, when the exposure has been severe, and long-continued, as in travelling, the disease has proved most difficult of removal.

Treatment.—The best course to pursue is, to keep such fowls in a warmed and dry domicile; the food being mild, yet nourishing: an occasional aperient of jalap should also be given. If the lameness does not seem to abate shortly, it will be well to give a pill, consisting of one grain of calomel, every other night: this will be found to give manifest relief.

ROUP.—This is one of the most serious diseases to which poultry are liable. It is, essentially, purulent catarrh, or catarrhal fever; the mucous membrane which lines the nasal passages being, in the more severe cases, affected with purulent inflammation (whilst in man it is the membranes of the throat, or the tonsils, which, in attacks of cold, scarlet fever, &c., are implicated in inflammation or ulceration).

Although the general febrile symptoms are not so discernible in fowls as in ourselves, yet the frequent thirst, loss of appetite, and generally enfeebled condition, as well as rapid wasting of the flesh, sufficiently indicate the presence of fever. That Roup is simply purulent catarrh, is evinced from the fact that, in the poultry-yard, where the disease is prevalent, it may be observed in all its stages of severity or mildness. In some fowls, the disease never amounts to more than a watery state of the eyes and nostrils : " but when," says Moubray, "the malady becomes confirmed, with running at the nostrils, swollen eyes, and other well-known symptoms, they are termed Roupy." In such examples, the discharge from the head is purulent, and peculiarly offensive, the constitutional disturbance being in ratio of severity with the local ones.

Symptoms.—When a fowl is Roupy, it begins to mope, to refuse its food, and is observed to resort frequently to the drinking vessel ; the eyes have a dull, heavy, and sunken appearance, while a frothy secretion sponges into their inner corners ; shortly, according to the severity of the attack, this becomes purulent ; whilst purulent matter is abundantly discharged from the nostrils. It is peculiarly offensive, and by concreting, closes up the eyelids as well as the nasal passages ; and now it not unfrequently is formed into sacs or abscesses on the cheeks. The eyelids are much swollen, and the head is hot and tumefied ; the breathing becomes difficult through the half-opened bill, as the natural air passage in the nostrils is occluded ; and the poor bird presents altogether a sad spectacle of wretchedness.

When Roup advances to the severe purulent stage, it is, for the most part, fatal ; and, even if the bird do not die, yet its recovery is always exceedingly slow and tedious. When the attack is less severe, by proper attention and good nursing, fowls are brought round without much difficulty.

Causes.—Cold, and especially when it is combined with damp or wet, is the exciting cause of Roup : unquestionably, however, there are predisposing causes, which render poultry more liable to become affected with the disease ; such are, want of cleanliness, insufficient or unwholesome food and water, confinement and impure air, &c. Such indeed are the baneful effects of these exciting and predisposing causes, that Roup may, at any time, be induced in fowls, if submitted to their combined agency.

Treatment.—As might be supposed, Roup has afforded a wide field for quackery: and numberless are the specifics which have from time to time been put forth, and by which the public have been largely deceived. In the treatment about to be recommended it will be observed, that medicines form but a part, and by no means the chief or most important part, of the remedial measures that are required.

In detailing the rational treatment of this, or indeed of any other disease, it is our duty to consider, first, what are the causes, both exciting and predisposing, of the complaint; and what are the nature and effects of such causes, either probable or ascertained, upon the body: secondly, what is the nature of the symptoms, or the character of the disease itself so induced; whether it be attended with high action or excitement, or with diminished power and exhaustion of the vital energies. Roup is eminently a disease of a debilitating type or nature, and marked by lingering exhaustion; though a degree of feverish excitement attends its first stage, yet this, like to the influenza in man, is attended with debility.

From the foregoing observations it will be anticipated, that the remedial measures about to be recommended will be of a tonic and restorative kind: the first and most essential step to be taken is, the removal of the bird from the sphere of those causes which produced the disease; a warmed, dry, yet sufficiently airy, domicile, forms indeed a most important part of successful treatment. Great attention should also be given to diet. The food must be nourishing and of easy digestion, and occasionally made more supporting by the addition of stimulants. To this end, barley-meal rather freely dredged with black pepper, and made into a stiff crumbly dough, is very appropriate. Bread sopped in ale may also occasionally be given, when the debility is great. A little wheat will sometimes be much relished as food, and will induce the fowl to partake more freely, where barley has been the usual grain allowed. Green food, which is too seldom included in the dietary of sick fowls, must be supplied. Chickweed (most aptly so called) is more relished than vegetables: this, or the leaves of lettuce as the best substitute, or even of broccoli, should be mixed freely with the food given.

As the bird is at first feverish, it will be of much service to give an occasional aperient. Though jalap is generally the best purgative that we possess for poultry, yet in this case, because of the great debilitating nature of the disease, it will be well to combine it with the leaves of rue, cut or minced very small, and made into a bolus (by the aid of a little butter) the size of a filbert: equal parts of jalap and of rue, or of jalap and cayenne capsules, may be used. Rue alone, though looked slightingly upon by many, is a very appropriate and good medicine in many diseases of fowls: its action is that of a stimulant aperient, and hence very appropriate in this disease. In smaller doses, about the size of a horse-bean, it is simply stimulant; this property being of course resident in the essential oil it contains. The small cayenne capsules or pods, with their contained seeds, as supplied by grocers, being crumbled small with the fingers, and made into a bolus the size of a horse-bean or hazel-nut, should be given night and morning. Powdered cubebs, given in the same way, but in larger doses, is of service; but much less efficient than the cayenne. The *essential oils* of copaiba and cubebs, in doses of ten drops, are excellent, but very unmanageable, and of most disagreeable and persistent smell.

It is certainly ascertained that these aromatic peppers have a great and almost specific influence over some purulent discharges from mucous membranes in man; and whether it be owing to the exertion of such specific influence on the purulent discharge of Roup, or whether because of their general stimulant and tonic effects, or both combined, yet of their peculiar efficacy in Roup there is no question whatever; indeed, the occasional use of jalap and rue, jalap and cayenne, and the cayenne boluses twice daily, constitute the most efficient medical treatment. In respect to local applications, it must be borne in mind, that the purulent discharge from the nostrils in Roup, like the ulcerated sore throat in scarlet fever, is but the local manifestation of constitutional disorder—a superadded symptom, marking the severity of the attack; and although oft becoming, by its long continuance and excess, itself a source of debility and of danger, yet it is an affection that is chiefly to be alleviated by means which restore the constitution itself; hence the acknowledged inefficacy of merely local means to remove the disorder.

By attention, however, to local applications, much discomfort and suffering may be saved to the fowl, and some amount of relief afforded to the violence of the symptoms. Frequent ablution of the eyelids, nostrils, and cheeks with warm water, in which sugar of lead is dissolved, in the proportion of two drachms to half a pint of rain-water, will be serviceable in lessening the swelling, as well as removing from these parts the copiously secreted, and often hardened, matter.

RUMP GLAND, *disease of.*—This oil-secreting gland sometimes becomes so obstructed and inflamed that suppuration takes place. The causes are by no means obvious, but the treatment is simple: it should be freely opened with a lancet, the contained matter gently pressed out, and, if it can be neatly done, well fomented, by squeezing a stream of hot water upon it from a spunge. The fowl must then be kept comfortable, and fed with mild soft food—an occasional aperient will also be very serviceable.

SOFT EGGS.—(See *Oviaries.*)

SPINAL DISEASE.—Some varieties of poultry are, apparently, more liable to Rickets, or distortion of the spine, than others. It is unquestionable that Polish fowls, for example, are especially subject to it. It is highly probable, however, that this peculiar disposition to the disease, is dependent upon in-breeding; the finer spangled varieties having, till very recently, been rare, and in the possession of but a few amateurs.

Symptoms.—The disease commences early; generally, when the chickens are but two or three months old. Sometimes the first symptom is, weakness of the legs, which causes them frequently to sit dowu, even when pecking up their food. When this is observed, it is called by some, simply leg-weakness, but the seat of the complaint is really in the spine. On cautiously examining the back, both with the eyes and the fingers, one thigh and hip will be found higher, or more prominent, than the other; depending upon lateral curvature of the spine itself. This state is never recovered from; but, as the bird grows older, it generally becomes more apparent. The more decided and confirmed symptom of this disease is, the tail being carried on one side, as if it had been twisted at the stump: there is also prominence, or a rounded appearance, of the back-bone.

Causes and Treatment.—The disease is evidently of a constitutional character, and most frequently induced by in-breeding. Any other combination of debilitating influences in the parent birds, as confinement, unwholesome or insufficient food, &c. &c., will also contribute to the production of diseased offspring. It is quite hopeless to attempt a cure. If there be simply a rounded appearance of the back, the fowl may be spared; but if the curvature of the spine be lateral, to one side, it should be at once destroyed.

CONCLUSION.—In addition to the diseases which have now been considered, dissection proves that poultry frequently also die of others; such as, tubercles in the lungs, deposition of tubercular matter, or the formation of fatty tumours on the liver, and other parts of the abdomen; chronic inflammation with thickening of the crop and stomach, and also acute inflammation of the bowels, or of the peritoneum, (that is, the external covering of the bowels and lining of the abdominal cavity). There are no diagnostic or distinguishing symptoms however on which we can rely, that indicate the presence of these various diseases; and they are, moreover, of so fatal a nature, that, even if we could detect them, we know of no medicine, at least, that would really remove them.

When fowls are afflicted with these chronic diseases, they show evident signs of being out of health, and begin to mope and pine away in flesh. When there is no ostensible cause for this, we may indeed suspect that tubercular deposition is taking place within the body, but nothing more.—In such cases we shall do well to let the bird have sufficient liberty; see that warmth and cleanliness are insured in the roosting-house; and, what is of especial moment, that they have *variety* of wholesome food, including *green* diet, fresh water, &c. An alterative of five grains of the *Hydr. cum Creta*, every other night, may prove serviceable, if we think tubercular deposition is threatened or commencing; and an occasional mild aperient dose of jalap with cayenne pepper, by promoting a healthy action of the digestive system, will afford the best chance of amendment.

On concluding these observations, the writer begs to remark that it was not at first contemplated to compose a treatise on the diseases of poultry, but simply to furnish to the able editor of this volume some practical hints, the result of his observation and experience. He soon found, however, that he must either "drink deep or taste not," the subject became more and more interesting—the statement of facts led to explanations of them—the recounting of symptoms and phenomena of diseases, irresistibly called forth reasoning and investigation of their cause and origin—so that at length the observations took the form in which they have now been presented.

So much the author vouches for—that nothing has been advanced but what was the well-considered and closely-tested results of personal investigation and reflection; the published statements and opinions of others were respectfully laid aside; and the writer observed, experimented, and thought for himself. So far from deeming the subject puerile, or unworthy of occupying serious attention, he has

found in it matter of deep and varied interest to the naturalist, the comparative anatomist, and the medical observer.
It would have been wholly inconsistent with the aim and
purpose of this popular work, had the subject of the diseases
of poultry not been presented in a plain and practical form ;
but it will be observed that the writer has not contented
himself with a simple statement of facts, but has also given,
when necessary, explanations and reasons for all that is set
forth.

The administration of drugs, it will be seen, has been but
sparingly recommended, and then only those that have a
simple, certain, and well-ascertained effect ; and that the
greatest reliance has in all cases been placed on due regulation of the diet, cleanliness, ventilation, warmth and comfort, &c.; for it cannot be too often repeated, that as the
diseases of poultry are most commonly induced by error, in-
attention, and neglect in these essential particulars, so is the
due and complete observance of them all-important for the
restoration of health.

In respect to the power of drugs in healing the languish-
ing and disordered bird, it is now replied, that, though the
advocates of some novel medicines may confidently refer
to restored fowls as evidence of their efficacy, it is much to
be feared that such persons have not yet learned to distinguish, in every instance, between a *cure* and a *recovery*.

NOTE ON THE DERIVATION OF THE NAME "POLISH FOWL."

MUCH doubt and discredit have been cast upon Polish Fowls, and
their title to be considered a distinct species questioned or denied ;
for the very unscientific reason, that their origin was enveloped in
deep obscurity ; and that though called Polish Fowls, it could not
be shown, that Poland was their special habitat, or that it was from
thence that they were derived.

2 K

True it is that their history is not specially connected with that part of the world; and that their name of Polish has been assigned for scientific, not for territorial, reasons. It is customary, in respect to the varieties and species, and even genera, of both plants and animals, to bestow upon them names, because of certain ostensible similitudes, or characteristics, by which they can be likened, or compared, to other familiar things: the entire nomenclature, indeed, in the sciences of Botany and Natural History, is really so derived and constructed.

It will now be shown that the fowls under consideration have, in accordance with such usage, received the appellation they now bear. And, although names are often given from almost trivial or imaginary resemblances, yet, in the case of Polish Fowls, it will be found that the peculiar anatomical conformation of the skull—and which is altogether unique—as well as the more ostensible mass of feathers, or topknot, on the head, afford not only an appropriate, but a natural and scientific reason, for the name imposed upon them.

This peculiarity consists in a large and prominent development of the cranium, or skull-cap, which, on dissection, I discovered to be remarkably elevated and round; whilst, in *all* other fowls, the head is flattened. In Polish chickens, this elevated and globular shape of the skull is so remarkable, that the brain appears as if encased in a round ball, lying on the top of the head. I conceive, then, that it is this singular and unique elevation of the top of the head, or *poll*, together, it may be added, with the ball of feathers thereon, that has furnished their characteristic cognomen of *Poll*-ish Fowl—that is, fowls with the peculiarly globular, elevated, and tufted *poll*—and which term has been naturally converted into *Poll*-ish. The word Poland, sometimes made use of, is evidently nothing more than a conversion of Polish; it is used synonymously, but Polish Fowl, not Poland Fowl, is the evident grammatical term, whether the birds came from Poland, or were named from personal characteristics.

This explanation of their name, in strict accordance with scientific usage, and natural as it is simple, at once frees us from all the speculation and fanciful opinion in which writers have indulged, when considering the history, or probable origin, of these fowls. Whilst this unique anatomical configuration of the skull suggests additional claim to their being considered a recognised and genuine species: and although it is true that it has not been before noticed by authors of our own times, yet it could not have always escaped the keen search of the older naturalists.

                                                      F. R. HORNER.

DISEASES OF PIGEONS.

CANKER. (See *Wounds.*)

CORE.—This is the name given to a complaint consisting of a hard, cheesy substance, of a yellowish white colour, and resembling the core of an apple, which is sometimes found in the vent, and also in the gullet. This will ripen and maturate, and may be then discharged, dissected, or drawn out. A purge of jalap may also be given.

CROP-BOUND, resulting from an obstruction of the food, generally from the pigeon having gorged too much. —This disorder is by far the most frequent in the Pouter pigeon, because of the large size of the crop. The distention is often exceedingly large, the crop sagging almost to the very ground, also being icy cold to the touch. The remedy is simple: place the bird in a warm or woollen stocking, the head and crop being supported as upright as may be; then hang the stocking up (the head of the bird upwards) in a warm room, for ten or twelve hours, when it will invariably be found that the food has passed the crop.

DIARRHŒA, *or* SCOURING.—A complaint too well known to need description.—The remedy, *chalk.* Give a bolus of compound chalk, the size of a bean, twice a day, till the bird is cured.

ERUPTIONS upon the skin, (perhaps from giving too much hemp seed,) sometimes known as *Small Pox:*—even if unattended to, we have never seen any ill effects result to the young pigeons, which are more usually attacked; and it seems to us to be merely a natural effort to throw peccant humours out of the system, which will generally disappear in a week or ten days. All external applications, though

2 K 2

generally recommended, are as useless and uncalled for, as in small pox in our own species.

GIDDINESS.—(See *Vertigo*.)

GIZZARD (FALLING).—This and many other peculiar internal diseases young pigeons are more particularly liable to; and are mostly caused by constitutional weakness or disorder. After numerous experiments, we are convinced that in all *confirmed* cases, (which may easily be known, by the lower part of the abdomen sagging on the ground,) the shortest way is to wring the necks of the poor birds— as they have never recovered under any treatment. It will be found that this large discoloured protrusion of the body, is really caused not by the simple falling of the gizzard, but by its being, with other internal parts, *propelled* downwards by hardened fatty growth from the liver, or upper part of the gizzard itself.

MEGRIMS. (See *Vertigo*.)

MOULTING.—It will sometimes happen that pigeons suffer much, and are unable to get through so well or quickly as they should; in which case resorting to high feeding on corn and hemp seed, with warmth, will greatly facilitate moulting. A little salt and alum constantly given dissolved in the water is also good at such times.

VERMIN.—These are generally the consequence of low keep and filth. Various remedies are offered, such as fumigating with spirits of turpentine, tobacco, snuff, and stavesacre; but the only effectual remedy is thorough and constant cleanliness, and providing, daily, pans of water for the pigeons to bathe in. If the bird is much pestered with vermin, they may be at once cleared off by thorough washing with soft soap and water, taking care, of course, to dry the feathers afterwards with a soft cloth and the warmth of a fire. A little mercurial ointment also, rubbed under the wing, on the sides, will kill the vermin.

VERTIGO.—This disease (but little understood by Fanciers generally) is closely allied to the apoplectic seizures in poultry, and depends on a similar state of the brain,—viz. congestion of the blood vessels, and finally effusion of blood thereon. The only beneficial treatment is, confinement to the loft, bleeding at the vein under the wing, and administering a few grains of jalap, or a comp. rhubarb pill.

WENS, or fleshy tumours, commonly on the shoulder-joint of the wings, not unfrequently occur in *old* birds which have been much *flown*. They are seldom cured, but may from time to time be cut off or opened, and washed with a solution of alum and water, or, which is preferable, have some dry powdered alum dusted over the parts.

WORMS, about 1½ inch long and ¼ inch broad, sometimes collect in a lump at the orifice of the vent; and can only be got rid of by injections of sweet oil and spirits of turpentine, continued for two or three days, but not oftener than once a day: or a compound rhubarb pill may be taken occasionally.

WOUNDS upon the head or wattles of Carriers and Barbes from fighting, should be carefully washed with a solution of sulphate of zinc and water, and anoint with a few drops of olive oil every day until well. If, however, the parts *canker*, as it is called, various remedies are adopted: the following will be found the most efficacious,—wash with two drachms of alum in one and half ounces of water or weak vinegar, or any spirit and water, and then anoint with an unguent composed of honey and powdered burnt alum; or mix twenty grains of red precipitate with half an ounce of honey, and use as the previous unguent; or dissolve five grains of white vitriol in half a table-spoonful of vinegar, and mix with alum and honey, and use as the other remedies.

WOUNDS on the feet or legs will generally heal better and sooner when let alone.

INDEX.

JOHN CHILDS AND SON, BUNGAY.

www.ingramcontent.com/pod-product-compliance
Lightning Source LLC
Chambersburg PA
CBHW080822220526
45467CB00008B/2171